INTERNATIONAL UNION OF CRYSTALLOGRAPHY
MONOGRAPHS ON CRYSTALLOGRAPHY

INTERNATIONAL UNION OF CRYSTALLOGRAPHY BOOK SERIES

This volume forms part of a series of books sponsored by the International Union of Crystallography (IUCr) and published by Oxford University Press. There are three IUCr series: IUCr Monographs on Crystallography, which are in-depth expositions of specialized topics in crystallography; IUCr Texts on Crystallography, which are more general works intended to make crystallographic insights available to a wider audience than the community of crystallographers themselves; and IUCr Crystallographic Symposia, which are essentially the edited proceedings of workshops or similar meetings supported by the IUCr.

IUCr Monographs on Crystallography

1 *Accurate molecular structures: Their determination and importance*
A. Domenicano and I. Hargittai, *editors*

2 *P. P. Ewald and his dynamical theory of X-ray diffraction*
D. W. J. Cruickshank, H. J. Juretschke, and N. Kato, *editors*

3 *Electron diffraction techniques, Volume 1*
J. M. Cowley, *editor*

4 *Electron diffraction techniques, Volume 2*
J. M. Cowley, *editor*

5 *The Rietveld method*
R. A. Young, *editor*

6 *Introduction to crystallographic statistics*
U. Shmueli and G. H. Weiss

IUCr Texts on Crystallography

1 *The solid state: From superconductors to superalloys*
A. Guinier and R. Jullien, *translated by* W. J. Duffin

2 *Fundamentals of crystallography*
C. Giacovazzo, *editor*

IUCr Crystallographic Symposia

1 *Patterson and Pattersons: Fifty years of the Patterson function*
J. P. Glusker, B. K. Patterson, and M. Rossi, *editors*

2 *Molecular structure: Chemical reactivity and biological activity*
J. J. Stezowski, J. Huang, and M. Shao, *editors*

3 *Crystallographic computing 4: Techniques and new technologies*
N. W. Isaacs and M. R. Taylor, *editors*

4 *Organic crystal chemistry*
J. Garbarczyk and D. W. Jones, *editors*

5 *Crystallographic computing 5: From chemistry to biology*
D. Moras, A. D. Podjarny, and J. C. Thierry, *editors*

6 *Crystallographic computing 6: A window on modern crystallography*
H. D. Flack, L. Parkanyi, and K. Simon, *editors*

7 *Correlations, transformations, and interactions in organic crystal chemistry*
D. W. Jones and A. Katrusiak, *editors*

Introduction to Crystallographic Statistics

URI SHMUELI
Professor of Chemistry
Tel Aviv University

GEORGE H. WEISS
Chief, Physical Sciences Laboratory
Division of Computer Research and Technology
National Institutes of Health

INTERNATIONAL UNION OF CRYSTALLOGRAPHY
OXFORD UNIVERSITY PRESS
1995

Oxford University Press, Walton Street, Oxford OX2 6DP
Oxford New York
Athens Auckland Bangkok Bombay
Calcutta Cape Town Dar es Salaam Delhi
Florence Hong Kong Istanbul Karachi
Kuala Lumpur Madras Madrid Melbourne
Mexico City Nairobi Paris Singapore
Taipei Tokyo Toronto
and associated companies in
Berlin Ibadan

Oxford is a trade mark of Oxford University Press

Published in the United States
by Oxford University Press Inc., New York

A catalogue record for this book is available from the British Library

Library of Congress Cataloging in Publication Data
(Data available)
ISBN 0 19 855926 7

Typeset by the authors using LaTeX
Printed in Great Britain by Bookcraft Ltd, Midsomer Norton

PREFACE

Modern applications of crystallography to the determination of crystal and molecular structure make very extensive use of statistical methodology. The resolution of space-group ambiguities, caused by an approximate validity of Friedel's law, is generally related to the probability density function (hereafter: pdf) of the magnitude of the structure factor. Direct methods of phase determination have been responsible for much of the success of crystallography. These methods are based on properties of joint pdfs that describe structure-invariant combinations of phases, conditioned on measured magnitudes of the corresponding structure factors. The importance of these aspects of structure-factor statistics is well appreciated by all crystallographers. However, the actual developments are confined to concise articles in the professional literature which, more often than not, presume a working knowledge of the theory of probability on the part of the reader. With these ideas in mind, and after having contributed to the development of structure-factor statistics, we decided to present the reader with a survey of techniques and theories in this field of research and, at the same time, to make this introductory survey as self-contained as possible.

Chronologically, the development of these probabilitistic approaches started with low-order approximations as introduced by Professor A.J.C. Wilson FRS in his famous 1949 article. This development has then proceded to higher approximations as expansions in terms of orthogonal polynomials, and has subsequently reached exact representations of the probability measures in terms of Fourier series. The literature on this subject is quite voluminous and no modern text seems to be available which introduces the reader to these different aspects of structure-factor statistics, while also presenting the probabilistic principles involved. The purpose of this monograph is to fill this gap.

The basic idea behind the book was formulated at Tel Aviv University and the National Institutes of Health over a period of several years. Most of the writing took place during the visit of one of us (U.S.) to the National Institutes of Health where, as before, a good collaboration between the authors was enjoyed. We are grateful to our wives, Miriam Shmueli and Delia Weiss, for their patience and encouragement. We would also like to express our sincere thanks to all our collaborators in the research pertaining to this monograph. These thanks are due to Mr James Kiefer, Ms Yael Posner, Dr Savely Rabinovich, Mr Ronen Shashua, Mr Richard Shrager, Ms Zafra Stein and Professor A.J.C. Wilson FRS whose collaboration we have greatly enjoyed.

Uri Shmueli
George Weiss

CONTENTS

1

INTRODUCTORY MATERIAL

1.1 Purpose and scope of the book

The determination of crystal and molecular structures achieved its present respectable status thanks to advances in techniques of data collection, the widespread availability of computing power and the development of methods for the interpretation of experimental data in terms of crystal symmetry and atomic and molecular models. Some of the most successful of these rely heavily on statistical methodology. Structure-factor statistics play an important role among these methods and many investigations of this subject can be found in the crystallographic literature, since its introduction by Wilson (1949). Early work on statistical properties of a single structure factor (or normalized intensity), as applied to intensity statistics, is summarized in the book of Srinivasan and Parthasarathy (1976). This work is mainly based on central limit theorem methods. A detailed treatment of approximate joint probability density functions (hereafter: pdfs) of several structure factors, relevant to the implementation of direct methods of phase determination, is given in a book by Giacovazzo (1980), which is also a good source of references to earlier work on the subject. There is no book or monograph which develops or summarizes exact expressions for the pdfs required to implement structure-factor statistics, and certainly none that addresses numerical problems in evaluating such expressions. Our main purpose is to develop pdfs of the magnitude of the structure factor. In addition, we discuss some of the approximations that result in less accurate but more conveniently computable expressions for these densities. In many instances we recall the mathematical tools that may be required by the reader for following details of the derivations. The quantities to be discussed in this monograph are of importance to the theories and practice of intensity statistics and direct methods.

We wish to recall that the fundamentals of crystallography are based, among other things, on a highly ordered arrangement of material units in the crystal, and a fully predictable geometry of the diffraction pattern that can be experimentally recorded. This high degree of order requires justification of the use of probabilistic methods. We have further thought it useful to provide the reader with an elementary background of the theory of probability, oriented towards its applications to the topics discussed in later chapters. This background material is included in Chapter 1, along with the representations of the structure factor to be used in the monograph.

Chapter 2 presents detailed derivations of pdfs of the magnitude $|E|$ of the normalized structure factor in their classical exact forms, as well as their limiting forms in the many-equal-atoms approximation. This chapter is intended to provide instructive illustrations of our methodology by simple and yet non-trivial examples. The connection of crystallographic problems with the random walk theory is emphasized throughout the chapter.

Chapter 3 introduces the reader to the central limit theorem of the theory of probability, which has been widely applied in crystallographic analyses. A proof of a simple but important version of this theorem is presented, and the theorem is applied to a rederivation of zeroth-order approximations to pdfs of $|E|$ for several situations of interest. We rederive the classical Wilson (1949) pdfs of $|E|$, proceed to the treatment of effects of non-crystallographic centrosymmetry, and treat the problem of correlated sums of random variables. These are exemplified by the pdfs of $|E|$ in the presence of dispersive scatterers in centrosymmetric or partial non-crystallographic centrosymmetry in non-centrosymmetric space groups, in the central limit theorem approximation. Chapter 3 concludes with a discussion of the lowest approximation to the conditional joint probability function of a three-phase invariant in space group $P1$.

Chapter 4 first discusses circumstances under which the central limit theorem gives poor results, and introduces higher-order approximations which are derived in terms of expansions in series of orthogonal polynomials. These expansions are then applied to the construction of generalizations of several pdfs based on the central limit theorem and discussed in Chapter 3. Chapter 4 also contains a detailed introductory discussion of moments of the trigonometric structure factor, and their application to the formation of exact even moments of $|E|$, required for the construction of the higher-order approximations discussed in this chapter.

In spite of the improvement offered over methods based on the central limit theorem, expansions in terms of orthogonal polynomials discussed in Chapter 4 not infrequently fail to furnish satisfactory approximation to the true pdf of $|E|$. This is due to the truncation of these expansions after a few terms which, in its turn, is caused by the rapidly increasing complexity of calculating successive terms and the difficulty in calculating higher moments of $|E|$ required for such extended expansions. Many of these difficulties can be avoided by the use of Fourier series expansions of pdfs of the structure factor in crystallographic applications. These Fourier series have coefficients of the same functional form and in many cases are quite simple to compute. The pdfs so expanded are of an '*ab initio*' type, insofar as they are not based on any previous (lower) approximations. They are mathematically exact and can be computed with little effort. This method is described in Chapter 5, starting from a detailed example. We then proceed to the introduction of the dependence of these pdfs on crystallographic symmetry, examine the effects of non-crystallographic symmetry and conclude with effects of the presence of heavy scatterers in special positions, and of dispersion. Chapter 5 deals with pdfs of the magnitude of a single structure factor, for arbitrary atomic composition, crystallographic symmetry and other factors mentioned above. It is independent of Chapters 3 and 4. However, representations of pdfs in terms of Fourier series can be shown to reduce to those predicted by the central limit theorem when the unit cell is composed of a sufficiently large number of identical atoms.

The last chapter of the monograph deals with some conditional joint pdfs of several structure factors. These functions are used in direct methods of phase determination, and are expressed in terms of exact Fourier series in Chapter 6. We first treat the conditional probability for the Σ_1 relationship in space group $P\bar{1}$, then proceed to derive the conditional probability for a positive sign of the important centrosymmetric structure invariant, $E(\mathbf{h})E(\mathbf{k})E(\mathbf{h}+\mathbf{k})$, and finally attack the problem of finding the exact conditional pdf of a three-phase invariant in space groups $P1$ and $P2$. We conclude the chapter, and the monograph, with a simple-to-evaluate parametric approximation of this density function, the parameters being based on exact representations of the conditional probability of the three-phase invariant. The methodology of Chapter 6 is analogous to that used in Chapter 5, the only difference being the number of structure factors (and number of Fourier summations) involved.

Our monograph presents applications of the theory of probability to structure-factor statistics starting from the lowest approximations but attempts to find algorithms that are both exact and hold good promise of being easy to evaluate numerically.

1.2 Representations of the structure factor

As mentioned earlier, we seek to establish in this monograph a number of probabilistic properties of structure factors, either singly or in groups. It therefore seems desirable to comment on the various representations of the structure factor that have been used in specific applications.

The conventional representation of the structure factor, in space group $P1$, is

$$F(\mathbf{h}) = \sum_{j=1}^{N} f_j \exp(2\pi i \mathbf{h}^T \mathbf{r}_j), \tag{1.2.1}$$

where N is the number of atoms in the unit cell, $\mathbf{h}$ is the diffraction vector, $\mathbf{r}_j$ is the position vector of the jth atom and f_j is the scattering factor of this atom. The scattering factor in eqn(1.2.1) depends both on the scattering angle and on the vibration parameters of the jth atom.

The structure-factor form given by eqn(1.2.1) is used in practice in calculations involving refinement of the atomic parameters but it is not the most convenient one for use in structure-factor statistics. This is so because of the angular dependence of the atomic scattering factors which gives rise to problems in various averaging steps. Thus, at least in the case of X-ray diffraction, we seek a representation in which this angular dependence, due to the temperature factor as well as to the transform of the electron density of an atom at rest (e.g. Coppens 1993), is eliminated insofar as possible. One such representation is afforded by defining the unitary structure factor

$$U(\mathbf{h}) = \sum_{j=1}^{N} u_j \exp(2\pi i \mathbf{h}^T \mathbf{r}_j), \tag{1.2.2}$$

where

$$u_j = f_j \left(\sum_{k=1}^{N} f_k \right)^{-1} \tag{1.2.3}$$

is the unitary scattering factor. When all the atoms are identical, each of the u_js obviously reduces to $1/N$ and the angular dependence is eliminated. Of course, some minor angular dependence may persist when the atomic motion is strongly anisotropic.

The unitary structure factor as written in eqn(1.2.2) appears in almost all studies concerning inequalities [Harker and Kasper (1948) and subsequent publications; for an extensive list of references see Giacovazzo (1980)] and has also been used in early studies of direct methods of sign (e.g. Cochran and Woolfson 1955) and phase (e.g. Cochran 1955) determination. However, the unitary structure factor gave way to the normalized structure factor, first introduced by Hauptman and Karle (1953*b*), which turned out to be more convenient in practical applications. The normalized structure factor can be defined as

$$E(\mathbf{h}) = \sum_{j=1}^{N} n_j \exp(2\pi i \mathbf{h}^T \mathbf{r}_j), \tag{1.2.4}$$

where

$$n_j = f_j \left(\sum_{k=1}^{N} f_k^2 \right)^{-1/2} \tag{1.2.5}$$

is the normalized scattering factor. Similar to the case of the unitary scattering factor, when all the atoms are identical, the n_j are a function of N alone

$$n_j = \frac{f}{[(Nf^2)^{1/2}]} = \frac{1}{\sqrt{N}}. \tag{1.2.6}$$

Most of the results to be derived in this monograph will be written in terms of the normalized structure factor. Note that these representations are applicable to general reflections only and the intensity average multiple must be taken into account when special reflections are considered (e.g. Giacovazzo 1980).

When the chemical composition of the asymmetric unit is heterogeneous, the unitary and normalized scattering factors do not depend only on N. However, their dependence on the scattering angle is weak and they can, for most practical purposes, be taken as constants which depend only on the atomic species.

The structure-factor representations in eqns (1.2.1), (1.2.2) and (1.2.4) have the general form

$$S(\mathbf{h}) = \sum_{j=1}^{N} s_j \exp(2\pi i \mathbf{h}^T \mathbf{r}_j), \tag{1.2.7}$$

which can be used for an explicit introduction of space-group symmetry. For example, if the crystal is centrosymmetric and the unit-cell origin is chosen at a crystallographic centre of symmetry, the position vectors $\mathbf{r}_j$ and $-\mathbf{r}_j$ refer to equivalent atoms. Equation (1.2.7) can then be written as

$$\begin{aligned} S(\mathbf{h}) &= \sum_{j=1}^{N/2} s_j \left[\exp(2\pi i \mathbf{h}^T \mathbf{r}_j) + \exp(-2\pi i \mathbf{h}^T \mathbf{r}_j)\right] \\ &= 2\sum_{j=1}^{N/2} s_j \cos(2\pi \mathbf{h}^T \mathbf{r}_j). \end{aligned} \tag{1.2.8}$$

Equation (1.2.8) corresponds to the space group $P\bar{1}$, provided no additional symmetry elements are present.

In the more general case the functional form of the structure factor is determined by the space group, by the choice of the space-group origin, and by the distribution of the atoms among the Wyckoff positions specific to the space group. For greater generality one should also consider anomalous dispersion and non-crystallographic symmetry. In what follows we adopt the space-group representation given in the *International Tables* (1983) but confine ourselves to the 'Origin 2' choices when a centrosymmetric space group is given in the Tables for two alternative choices of the origin.

Let the asymmetric unit contain M atoms, which are distributed among k types of Wyckoff positions so that

$$M = M_1 + M_2 + \cdots + M_k, \tag{1.2.9}$$

where M_m is the number of atoms contributed to the asymmetric unit by the mth Wyckoff position. If, for example, an atom is located on an n-fold axis of rotation, only $1/n$ of the atomic volume is contributed to the asymmetric unit and summation over fractions may therefore be necessary. It is therefore convenient to assign to each Wyckoff position an occupancy factor, say a_m, which equals the fraction of the atomic volume contributed by this position to the asymmetric unit of the space group. The structure factor then takes the form

$$S(\mathbf{h}) = \sum_{m=1}^{k} a_m \sum_{j=1}^{M_m} s_j [\xi_j^m(\mathbf{h}) + i\eta_j^m(\mathbf{h})], \tag{1.2.10}$$

where the outer summation in (1.2.10) ranges over the types of (occupied) Wyckoff positions, a_m is the (possibly fractional) occupancy factor of the mth Wyckoff position, the inner summation in (1.2.10) ranges over the atoms occupying this (mth) position and $\xi_j^m + i\eta_j^m$ is the complex trigonometric structure factor of atom j in the mth Wyckoff position. The real and imaginary parts of this trigonometric structure factor are given by

$$\xi_j^m(\mathbf{h}) = \sum_{s=1}^{g_m} \cos\left[2\pi \mathbf{h}^T(\mathbf{P}_s^m \mathbf{r}_j + \mathbf{t}_s)\right] \tag{1.2.11}$$

and

$$\eta_j^m(\mathbf{h}) = \sum_{s=1}^{g_m} \sin\left[2\pi \mathbf{h}^T(\mathbf{P}_s^m \mathbf{r}_j + \mathbf{t}_s)\right], \tag{1.2.12}$$

where the summation ranges over the space-group operations transforming a reference atom in Wyckoff position m to its equivalents within the unit cell, $(\mathbf{P}_s^m, \mathbf{t}_s)$, $s = 1, 2, \ldots, g_m$, are the corresponding matrix-and-vector representations of these operations (*International Tables* 1983) and $\mathbf{r}_j$ is the position vector of the jth atom. Simplified expressions for the trigonometric structure factors corresponding to general positions (i.e. to $a_m = 1$) are given in *International Tables* (1965, 1993) for all the three-dimensional space groups. There is no ordered catalogue of the corresponding expressions for special positions.

In most calculations involved in the structure-factor statistics treated in the present monograph, the simplified formulae for the trigonometric structure factors are indispensable. As a simple example let us consider those for the general positions of the space group $Pmm2$. The coordinates of the equivalent positions, i.e. $\mathbf{Pr}+\mathbf{t}$, for $Pmm2$ are given by: x, y, z; $\overline{x}, y, z$; $x, \overline{y}, z$; $\overline{x}, \overline{y}, z$. Substituting these values in eqns (1.2.11) and (1.2.12), and carrying out some simple algebra, we obtain

$$\xi(hkl) = 4\cos(2\pi hx)\cos(2\pi ky)\cos(2\pi lz) \tag{1.2.13}$$

and

$$\eta(hkl) = 4\cos(2\pi hx)\cos(2\pi ky)\sin(2\pi lz). \tag{1.2.14}$$

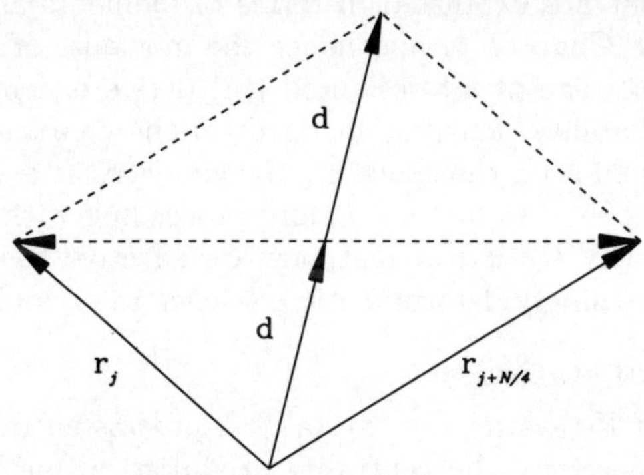

FIG. 1.1. *Geometry of non-crystallographic centrosymmetry.*

Modifications of the functional form of the structure factor are required to take into account the coexistence of crystallographic and non-crystallographic symmetry. Let us consider a crystal belonging to space group $P\overline{1}$, the asymmetric unit of which possesses a centre of symmetry not coinciding with a crystallographic centre. If we assume that all the atoms are located in general positions, the asymmetric subunit now contains $N/4$ atoms, where N is the number of atoms in the unit cell. If we denote the position vector of the non-crystallographic centre by $\mathbf{d}$ (see Fig. 1.1) it is readily shown that the additional centre relates the atoms of the asymmetric unit by the equation

$$\mathbf{r}_{j+N/4} = 2\mathbf{d} - \mathbf{r}_j. \tag{1.2.15}$$

The structure factor is therefore given by

$$\begin{aligned} F(\mathbf{h}) &= 2\sum_{j=1}^{N/4} f_j[\cos(2\pi\mathbf{h}^T\mathbf{r}_j) + \cos(2\pi\mathbf{h}^T\mathbf{r}_{j+N/4})] \\ &= 2\sum_{j=1}^{N/4} f_j\left\{\cos(2\pi\mathbf{h}^T\mathbf{r}_j) + \cos[2\pi\mathbf{h}^T(2\mathbf{d} - \mathbf{r}_j)]\right\} \\ &= 4\cos(2\pi\mathbf{h}^T\mathbf{d})\sum_{j=1}^{N/4} f_j\cos[2\pi\mathbf{h}^T(\mathbf{r}_j - \mathbf{d})]. \end{aligned} \tag{1.2.16}$$

Similar considerations lead to the forms of the structure factor, modified by the presence of non-crystallographic symmetry, in other space groups.

The structure-factor representations outlined above will be used in the present monograph to: (i) construct pdfs based on the central limit theorem, the so-called ideal pdfs (see Chapter 3), (ii) derive corrections to the ideal pdfs that are expressed in terms of moments and cumulants of this quantity (see Chapter 4) and hence the moments of the trigonometric structure factor are of interest, and (iii) derive exact formulations of the relevant probability densities in terms of the characteristic function. This function depends on the space-group symmetry as expressed in terms of trigonometric structure factors. Before proceeding with these topics we wish to explain why statistical methods are effective tools for analyzing problems in as seemingly deterministic a science as crystallography.

1.3 The role of statistics

The scattering of X-rays from a crystal is a process entirely described by physical laws. Therefore, the centrality of statistical methodology in the interpretation of crystallographic data is not an obvious idea. Indeed, the application of probabilistic methods in crystallography only first appeared in the works of Wilson (1942, 1949) and shortly thereafter, and more extensively, in the work of Hauptman and Karle (1953*a*, and subsequent publications) and others who dealt with both intensity statistics and direct methods.

In order to understand why statistics plays a role in crystallography we return to section 1.2, in which the structure factor is expressed as a linear combination of trigonometric terms, the phase of the contribution from atom j in the unit cell being denoted by θ_j, where

$$\theta_j \equiv 2\pi \mathbf{h}^T \mathbf{r}_j = 2\pi(hx_j + ky_j + lz_j), \tag{1.3.17}$$

where h, k and l are integers and x_j, y_j and z_j are fractional atomic coordinates. Since the θ_j always appear as arguments of trigonometric functions we only need to know how the θ_j are distributed in the interval $(0,2\pi)$ or, equivalently, in $(-\pi, \pi)$ to describe the properties of the structure factor in probabilistic terms. This restriction is equivalent to our being interested in the sequence

$$\rho_j = hx_j + ky_j + lz_j - [hx_j + ky_j + lz_j] = \frac{\theta_j}{2\pi} - \left[\frac{\theta_j}{2\pi}\right] \tag{1.3.18}$$

where by '$[X]$' we will mean 'the largest integer contained in X'. Thus, for example, if $hx_j + ky_j + lz_j = 3.4$, the corresponding value of ρ_j is 0.4.

Two different physical pictures can be used to motivate our remarks: (i) the coordinates $x_j y_j z_j$ are fixed and each of h, k and l is allowed to vary over a large set of integers, and (ii) the components hkl are fixed and each of x_j, y_j and z_j takes on random values. We consider first the case in which the coordinates are fixed and the integers h, k and l are varied.

The collection of experimental information includes data taken at a large number of diffraction vectors $\mathbf{h}(hkl)$, which in turn generates a large number of values of θ_j. Rather remarkably, one can say something about the distribution of the set $\{\theta_j\}$, $j = 1, 2, 3, \ldots$, provided that the atomic coordinates have the mathematical property known as *rational independence.* By a set of rationally independent variables, (x, y, z), we will mean that there does not exist a set of integers (m_1, m_2, m_3, m_4) such that

$$m_1 x + m_2 y + m_3 z = m_4, \tag{1.3.19}$$

with the integers m_1, m_2, m_3 and m_4 not all equal to zero. When the atomic coordinates are rationally independent, a theorem first developed in the context of number theory (Bohl 1909; Sierpinski 1910; Weyl 1916) provides the key information relating to the distribution of the ρ_j defined in (1.3.18). The theorem states that if one considers the sequence $\rho_1, \rho_2, \ldots, \rho_n$, then, as n becomes very large, the ρ_j increasingly tend to have the properties of independent random variables that are uniformly distributed in the interval (0,1). That is to say, if one divides up the unit interval into a set of equal subintervals of width $1/m$, m being an integer, the number of ρs out of a very long sequence that fall within any given subinterval will tend towards the value n/m, n being the number of ρs. This striking result, the proof of which is outside the scope of this monograph (see, for example, Hardy and and Wright 1979), allows us to conveniently describe the distribution of the θ_j in probabilistic terms. In particular we can denote the pdf of the variable θ_j to be $p(\theta_j)$ where this function is to be interpreted in the following way: the probability that an arbitrary value of θ falls in the interval $(\theta, \theta + d\theta)$ is equal to $p(\theta)d\theta$, in the limit $d\theta \to 0$. By the theorem cited earlier, in the absence of rationally dependent atomic coordinates the function $p(\theta)$ is constant in the interval $(-\pi, \pi)$, that is

$$p(\theta) = \frac{1}{2\pi}, \quad -\pi < \theta \leq \pi. \tag{1.3.20}$$

Since we can describe the θs in probabilistic terms (technically, as random variables), we can bring a number of techniques from probability theory to bear on the solution of some important problems in crystal structure determination. Detailed applications of these techniques are given in the forthcoming chapters. We emphasize the importance of the assumption of rational independence as defined in eqn(1.3.19) above. Some examples of computer-generated sets of ρ_j are shown in Figs 1.2 and 1.3. The relationship of these results to the presence or absence of rational dependence and the number-theoretical theorem cited earlier are clearly illustrated in the figures.

Let us now consider the second situation, in which we fix the integers h, k and l and let each of the x_j, y_j and z_j be completely unknown. As will become clear from the next section, each of these fractional coordinates can be regarded as a random variable. Our initial ignorance of the structure

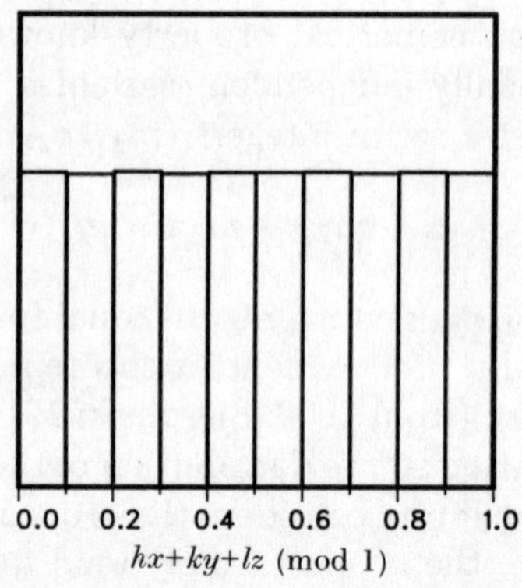

FIG. 1.2. *Distribution of $hx + ky + lz$ for weak rational dependence of x, y, z.*

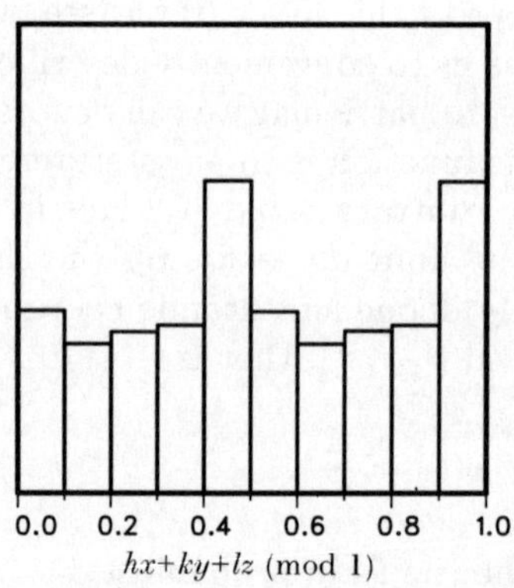

FIG. 1.3. *Distribution of $hx + ky + lz$ for strong rational dependence of x, y, z.*

implies a uniform probability for any of the points in the range of existence of each of these coordinates. In other words, each of x_j, y_j and z_j can be taken as being uniformly distributed on (0,1). By this very definition, the resulting ρs are also uniformly distributed on (0,1), i.e. the fractional part of the scalar product $\mathbf{h}^T\mathbf{r}_j$, which appears in all relevant formulations of the structure factor, can again be regarded as a uniformly distributed random variable.

It should be pointed out that the assumption of uniform distribution may not hold to the same extent for all crystalline substances. The arguments so far presented imply that an interatomic distance may be arbitrarily small and it is obvious that this cannot be the case. Excluded volume effects due to symmetry elements have been investigated by Nigam and Wilson (1980), but no general theory has so far been found applicable. However, the success of statistical methods in crystallography indicates that these stereochemical effects are probably rather small. An obvious case of predictable rational dependence is the presence of heavy scatterers in special positions of a space group. We shall derive pdfs in Chapter 5 which account for this situation.

1.4 Fundamental ideas in probability theory

Since the application of probability theory to crystallographic analysis is the central theme of this monograph it is appropriate to recall some of the basic ideas and definitions of the theory to be used in later chapters. A more complete and very readable introduction to probability theory can be found in the text by Feller (1971), but we will need much less of the theory than is covered in such a text.

The objects dealt with by probability theory are known as *random variables.* A rigorous definition of the term 'random variable' can lead to subtle mathematical and philosophical considerations, but we will use it in the most colloquial sense as meaning a variable whose value cannot be predicted in advance of an experiment. The state of the weather is an excellent example of such a variable, but also suggests some of the deeper problems that can be raised in framing the definition. If all of the dynamical factors that influence the weather were known then there would be no need to attach a probability to a forecast. In the absence of such comprehensive information, the meteorologist is forced to predict rain only in probabilistic terms. The reduction of complex dynamical systems like the weather to one in which probabilistic statements suffice for a description is not completely understood, nor is the reason why probability, a purely mathematical theory, should provide results that apply to so much of physical reality. In the most naive sense one can justify the application of probability theory to the physical sciences by the observation that it seems to work in a large number of situations. This is certainly also the case in crystallographic applications.

In section 1.2 we have seen that the normalized structure factor for a compound crystallizing in space group $P\bar{1}$ can be written as a sum of trigonometric terms, of which the ith term is $2n_i \cos\theta_i$, where $\theta_i = 2\pi\mathbf{h}^T\mathbf{r}_i$. The implication of the theorem cited in section 1.3 [often, but erroneously, attributed entirely to Weyl (1916)], under some often realized structural assumptions, is that the variable θ_i can be regarded as a uniformly distributed random variable whose pdf is uniform as shown in (1.3.20). Since a sum of random variables is itself a random variable the normalized structure factor

$$E(\mathbf{h}) = 2\sum_{j=1}^{N/2} n_j \cos\theta_j \tag{1.4.21}$$

can also be regarded as a random variable with its own associated pdf. Equation (1.3.20) is the appropriate pdf for any of the θ_i in the absence of rational dependence. We will later want to find the pdf of $E(\mathbf{h})$, given the pdfs of the θs, including its dependence on space-group symmetry as well as on other factors. However, before tackling this problem we wish to recall a certain amount of terminology as well as a number of functions to be used later in this monograph.

1.4.1 *The probability density function (pdf) and its moments*

Let X be a scalar random variable which can take on a continuum of values. The pdf of X, $p(x)$, is defined so that $p(x)dx$ is the probability that X falls in the interval $(x, x + dx)$. It is important to distinguish between X, the random variable whose value can only be estimated in terms of experimental data, and the variable x which occurs as an argument of $p(x)$ and is a specified parameter. When X has a dimension, say length, then $p(x)$ has the dimension $(\text{length})^{-1}$ since $p(x)dx$ is a probability which must itself be dimensionless. The definition of $p(x)$ as a density implies the obvious property

$$p(x) \geq 0. \tag{1.4.22}$$

The probability that X lies in a finite interval (a, b) is then

$$\Pr\{a < X \leq b\} = \int_a^b p(x)dx. \tag{1.4.23}$$

Since a probability equal to 1 is equivalent to knowledge with certainty, and the variable X must lie with certainty somewhere in the range $(-\infty, \infty)$, it follows that $p(x)$ must satisfy

$$\int_{-\infty}^{\infty} p(x)dx = 1. \tag{1.4.24}$$

The cumulative distribution, $P(x)$, is defined to be the probability that $X \leq x$ which, according to eqn(1.4.23) can be expressed in the form

$$P(x) = \int_{-\infty}^{x} p(u)du, \tag{1.4.25}$$

so that $P(-\infty) = 0$ and $P(\infty) = 1$. A useful set of parameters associated with a pdf or cumulative distribution function is that of the moments. The nth moment of X, $(n = 1, 2, 3, \ldots)$ will be denoted by the relation

$$\langle X^n \rangle = \int_{-\infty}^{\infty} x^n p(x) dx. \tag{1.4.26}$$

These will be said to exist when the indicated integral converges. In crystallographic applications all moments are finite although in other applications they may not be. While a given pdf uniquely determines its moments, it is not necessarily true that the set of all moments uniquely determines the form of $p(x)$ (Shohat and Tamarkin 1943). The first moment, $\langle X \rangle$, is known as the *mean* or the *average* value of X. A quantity related to the second moment is the *variance*, which is usually denoted by σ^2 and is defined by $\sigma^2 = \langle X^2 \rangle - \langle X \rangle^2$. This quantity is a measure of the dispersion in the pdf since σ^2 can also be defined by the relation

$$\sigma^2 = \langle (X - \langle X \rangle)^2 \rangle. \tag{1.4.27}$$

If X is identically equal to a constant (which is necessarily equal to $\langle X \rangle$) then $\sigma^2 = 0$, otherwise $\sigma^2 > 0$ since $(x - \langle x \rangle)^2$ is positive except at points at which $X = \langle X \rangle$.

1.4.2 *The characteristic function*

It is convenient, as will shortly be demonstrated, to introduce a function related to the pdf by a Fourier transform, known as the *characteristic function* which will be denoted by $C(\omega)$. In a later paragraph we shall see that the characteristic function has certain desirable features for analysing statistical properties of sums of independent random variables. The definition of the characteristic function is

$$C(\omega) = \langle \exp(i\omega x) \rangle = \int_{-\infty}^{\infty} \exp(i\omega x) p(x) dx. \tag{1.4.28}$$

The inverse of this relation allows one to express $p(x)$ in terms of $C(\omega)$. Since the characteristic function is a Fourier transform, its inverse is known to be

$$p(x) = \frac{1}{2\pi} \int_{-\infty}^{\infty} \exp(-i\omega x) C(\omega) d\omega \tag{1.4.29}$$

(e.g. Cramér 1946). Several properties of $C(\omega)$ are trivially obtained from the representation in eqn(1.4.28). By comparing the latter equation with eqn(1.4.24) one sees that

$$C(0) = 1. \tag{1.4.30}$$

A second important result implied by eqn(1.4.28) is that $|C(\omega)| \leq 1$ for all ω. This can be seen from the bounds

$$|C(\omega)| = \left| \int_{-\infty}^{\infty} \exp(i\omega x) p(x) dx \right| \leq \int_{\infty}^{\infty} |\exp(i\omega x)| p(x) dx$$

$$= \int_{-\infty}^{\infty} p(x)dx = 1. \qquad (1.4.31)$$

By suitably restricting the class of $p(x)$ one can show that $|C(\omega)|$ is strictly less than 1 when $\omega \neq 0$ (Gnedenko and Kolmogorov 1954). In crystallographic applications this will always be the case, which will later be of importance in our discussion of the central limit theorem.

A final important property of the characteristic function is found by formally expanding the exponential in the integrand of eqn(1.4.28) in a Taylor series. This expansion is

$$\begin{aligned} C(\omega) &= \sum_{n=0}^{\infty} \frac{i^n \omega^n}{n!} \int_{-\infty}^{\infty} x^n p(x)dx \\ &= \sum_{n=0}^{\infty} \frac{i^n \omega^n}{n!} \langle X^n \rangle \end{aligned} \qquad (1.4.32)$$

provided that all the moments are finite. From this equation it follows that moments can be found in terms of derivatives of $C(\omega)$ at $\omega = 0$. We have in fact the relations

$$\langle X^n \rangle = (-1)^n \left. \frac{d^n C(\omega)}{d\omega^n} \right|_{\omega=0} . \qquad (1.4.33)$$

1.4.3 *Cumulants and moments*

A set of parameters related to moments, but more convenient in certain applications, is the set of *cumulants*, $\{\kappa_n\}, n = 1, 2, 3, \ldots$. These are generated from the *cumulant generating function*, $K(\omega)$, which is derived from the characteristic function by

$$K(\omega) \equiv \sum_{n=1}^{\infty} \frac{i^n \kappa_n \omega^n}{n!} = \ln C(\omega). \qquad (1.4.34)$$

Since $C(\omega)$ can be written in terms of a series of moments as shown in eqn(1.4.32), and $K(\omega)$ is just a function of $C(\omega)$, it follows that the cumulants, κ_n, can be expressed as non-linear functions of the moments and vice versa. The first few expressions for cumulants in terms of moments are readily found to be

$$\begin{aligned} \kappa_1 &= \langle X \rangle \\ \kappa_2 &= \langle X^2 \rangle - \langle X \rangle^2 \\ \kappa_3 &= \langle X^3 \rangle - 3\langle X^2 \rangle \langle X \rangle + 2\langle X \rangle^2 \\ \kappa_4 &= \langle X^4 \rangle - 4\langle X^3 \rangle \langle X \rangle - 3\langle X^2 \rangle^2 + 12\langle X^2 \rangle \langle X \rangle^2 - 6\langle X \rangle^4 \end{aligned} \qquad (1.4.35)$$

and the corresponding expressions for moments in terms of cumulants are

$$\begin{aligned}\langle X\rangle &= \kappa_1\\ \langle X^2\rangle &= \kappa_2+\kappa_1^2\\ \langle X^3\rangle &= \kappa_3+3\kappa_1\kappa_2+\kappa_1^3\\ \langle X^4\rangle &= \kappa_4+4\kappa_3\kappa_1+3\kappa_2^2+6\kappa_2\kappa_1^2+\kappa_1^4.\end{aligned} \tag{1.4.36}$$

A more extensive list of these relations is found in the book by Stuart and Ord (1987), as are general formulae allowing one to transform moments to cumulants and vice versa.

1.4.4 *An example*

As a simple, but later useful, application of the theory outlined above we shall calculate the various parameters associated with the random variable $X = \cos\theta$ on the assumption that θ is uniformly distributed over $(-\pi, \pi)$, i.e. $p(\theta) = (2\pi)^{-1}$. One can write an expression for the moments given in terms of known integrals

$$\langle X^n\rangle = \begin{cases} \frac{1}{2\pi}\int_{-\pi}^{\pi}\cos^n\theta d\theta = \frac{n!}{2^n(n/2)!} & \text{if } n \text{ is even}\\ 0 & \text{if } n \text{ is odd.}\end{cases} \tag{1.4.37}$$

The characteristic function is readily evaluated as

$$C(\omega) = \frac{1}{2\pi}\int_{-\pi}^{\pi}\exp(i\omega\cos\theta)d\theta = J_0(\omega), \tag{1.4.38}$$

where $J_0(\omega)$ is a zeroth-order Bessel function (Gradshteyn and Ryzhik 1980). The cumulant generating function is given by the logarithm of $J_0(\omega)$. One can find an explicit expansion of $K(\omega)$ in this particular example (Kiefer and Weiss 1984):

$$K(\omega) = -\sum_{m=1}^{\infty}\alpha_m\frac{\omega^m}{m} \tag{1.4.39}$$

in which the α_m are constants, the first few of which are

$$\alpha_1 = \frac{1}{4},\ \alpha_2 = \frac{1}{32},\ \alpha_3 = \frac{1}{192},\ \alpha_4 = \frac{11}{12,288},\cdots. \tag{1.4.40}$$

The combination of eqns (1.4.34), (1.4.39) and (1.4.40) allows us to find the cumulants explicitly.

1.4.5 *Pdfs, characteristic functions and moments of vector random variables*

Crystallographic applications often require the analysis of combinations of random variables. For example, in order to find the pdf of the structure factor for a compound belonging to space group $P\bar{1}$ one has to analyse the properties of the sum $F(\mathbf{h}) = 2\sum_j f_j \cos\theta_j$, where each of the θ_js itself is a random variable. Many of the definitions given in the context of single random variables generalize in a trivial fashion to sets of variables, but there are some new features as well. The first function required in a number of crystallographic calculations is the *joint pdf*, $p(x_1, x_2, \ldots, x_n) \equiv p(\mathbf{r})$ of n random variables $X_1, X_2, \ldots, X_n$. Analogous to the definition of $p(x)$, the function $p(x_1, \ldots, x_n)dx_1 \cdots dx_n$ is the joint probability that: $x_1 < X_1 \leq x_1 + dx_1, x_2 < X_2 \leq x_2 + dx_2, \ldots, x_n < X_n \leq x_n + dx_n$. The normalization condition for $p(x)$ analogous to eqn(1.4.24) is

$$\int_{-\infty}^{\infty} \int_{-\infty}^{\infty} p(x_1, \ldots, x_n)dx_1 \cdots dx_n = 1. \qquad (1.4.41)$$

The marginal pdf of a subset $x_1, x_2, \ldots, x_m$ of the total set $x_1, x_2, \ldots, x_n$ is defined to be the pdf of the values of members in the subset when the remaining variables are allowed to take on all possible values. Let this pdf be denoted by $p_{marg}(x_1, x_2, \ldots, x_m)$, which by our definition can be expressed in terms of $p(x_1, x_2, \ldots, x_n)$ as

$$p_{marg}(x_1, \ldots, x_m) = \int_{-\infty}^{\infty} \cdots \int_{-\infty}^{\infty} p(x_1, \ldots, x_n)dx_{m+1}dx_{m+2} \cdots dx_n, \qquad (1.4.42)$$

where m can take on any value from 1 to n. In much of the following analysis it will be convenient to use the multivariate characteristic function $C(\omega_1, \omega_2, \ldots, \omega_n) \equiv C(\boldsymbol{\omega})$ which is the generalization of eqn(1.4.28)

$$C(\omega_1, \ldots, \omega_n) = \int_{-\infty}^{\infty} \cdots \int_{-\infty}^{\infty} p(x_1, \ldots, x_n) \exp\left(i\sum_{j=1}^{n} \omega_j x_j\right) \prod_{j=1}^{n} dx_j \qquad (1.4.43)$$

which can be written more succinctly in vector notation as

$$C(\boldsymbol{\omega}) = \int_{-\infty}^{\infty} \cdots \int_{-\infty}^{\infty} p(\mathbf{r}) \exp(i\boldsymbol{\omega}^T \mathbf{r})d^n\mathbf{r}, \qquad (1.4.44)$$

where $\mathbf{r} = (x_1, x_2, \ldots, x_n)$. The inverse of this relation is

$$p(\mathbf{r}) = \frac{1}{(2\pi)^n} \int_{-\infty}^{\infty} \cdots \int_{-\infty}^{\infty} C(\boldsymbol{\omega}) \exp(-i\boldsymbol{\omega}^T \mathbf{r})d^n\boldsymbol{\omega}. \qquad (1.4.45)$$

The average of a function, $f(\mathbf{r})$, will be denoted by $\langle f(\mathbf{r})\rangle$ *which is defined by*

$$\langle f(\mathbf{r})\rangle = \int_{-\infty}^{\infty} \cdots \int_{-\infty}^{\infty} f(\mathbf{r})p(\mathbf{r})d^n\mathbf{r}. \tag{1.4.46}$$

In particular, one can define the mixed moments with respect to $p(\mathbf{x})$ which are the averages $\langle X_1^{m_1} X_2^{m_2} \dots X_n^{m_n}\rangle$. The characteristic function can be expanded in a multiple Taylor series in terms of the mixed moments analogous to the expansion in eqn(1.4.32). Similarly, it is possible to define a cumulant generating function in the multivariate case by $K(\boldsymbol{\omega}) = \ln[C(\boldsymbol{\omega})]$.

1.4.6 *Conditional pdfs*

An extension of the notion of the joint pdf is that of the *conditional joint pdf.* Let us suppose that the set $\{X_i\}, i = 1, 2, 3, \dots, n$, consists of random variables, but let us also suppose that $X_m, X_{m+1}, \dots, X_n$ have been determined in some way so that they may be regarded as known quantities. This knowledge suffices to change the pdf of the remaining variables. As a trivial example of the distinction between a pdf and a conditional pdf, if I arrive at a bus stop to wait for a bus that comes every five minutes and do not know the time of the last bus arrival then I can calculate a pdf for my waiting time which will be non-zero at all times between zero and five minutes. On the other hand, if I know that the last bus arrived two minutes before my own arrival then I automatically know that there are exactly three minutes remaining to the arrival of the next bus. Thus, my knowledge of the arrival time has changed the form of the pdf of the remaining waiting time from being one which can take on values from zero to five minutes to a delta function form, i.e. one which differs from zero at a single time only.

To define the function which describes conditioning, we denote the pdf of the random variables $X_1, X_2, \dots, X_{m-1}$ — *given a knowledge* of a second set $X_m, X_{m+1}, \dots, X_n$ — by $p(x_1, \dots, x_{m-1}|x_m, \dots, x_n)$. To find an expression for the conditional pdf we note that the pdf of X can be regarded as the conditional pdf of the subset $\{X_1, X_2, \dots, X_{m-1}\}$ when the remaining variables are allowed to vary according to their appropriate marginal pdf. In other words

$$p(\mathbf{r}) = p(x_1, \dots, x_{m-1}|x_m, \dots, x_n)p_{marg}(x_m, \dots, x_n) \tag{1.4.47}$$

or

$$p(x_1, \dots, x_{m-1}|x_m, \dots, x_n) = \frac{p(x_1, \dots, x_n)}{\int_{-\infty}^{\infty} \cdots \int_{-\infty}^{\infty} p(x_1, \dots, x_n)dx_1 dx_2 \dots dx_{m-1}}. \tag{1.4.48}$$

Conditional pdfs are of central importance in multivariate structure-factor statistics, and are often applied in the theory and practice of direct methods.

1.4.7 *Independent random variables*

A central concept needed in the remainder of this monograph is that of *the independent random variables.* An intuitive definition of this idea can be summarized in the statement that two (or more) random variables are independent if they do not influence each other. For example, in a sequence of coin tosses it is generally believed that the outcome of a single toss does not influence the outcome of any subsequent toss. This is equivalent to the statement that the outcomes are independent. To put this notion into more mathematical terms we will say that the random variables $X_1, X_2, \ldots, X_n$ are independent when their joint pdf, $p(\mathbf{r})$, can be decomposed into a product of univariate pdfs

$$p(x_1, x_2, \ldots, x_n) = p(x_1)p(x_2)\cdots p(x_n). \tag{1.4.49}$$

For a set of independent random variables the mixed moments, i.e. the averages $\langle X_1^{m_1} X_2^{m_2} \cdots X_n^{m_n} \rangle$ can be factored as

$$\langle X_1^{m_1} X_2^{m_2} \cdots X_n^{m_n} \rangle = \langle X_1^{m_1} \rangle \langle X_2^{m_2} \rangle \cdots \langle X_n^{m_n} \rangle \tag{1.4.50}$$

which is evident on substituting eqn(1.4.49) into the definition of moments given in eqn(1.4.46). Notice also that for independent random variables the conditional pdfs satisfy

$$p(x_1, \ldots, x_{m-1} | x_m, \ldots, x_n) = \prod_{j=1}^{m-1} p(x_j). \tag{1.4.51}$$

That is to say, a knowledge of the values of any subset of variables cannot change our state of knowledge about the properties of the remaining variables.

1.4.8 *Characteristic functions for sums of independent random variables*

One final result that plays a significant role in the remainder of the monograph is the decomposition property of the characteristic function of a sum of independent random variables. For simplicity we consider the case in which the X_is are scalar variables. Let $Y = X_1 + X_2 + \cdots + X_n$, where $\{X_i\}$ is a set of independent random variables. The characteristic function of Y is

$$C(\omega) = \langle \exp[i\omega(X_1 + X_2 + \cdots + X_n)] \rangle. \tag{1.4.52}$$

However, because $p(x)$ has the decomposition property shown in eqn (1.4.51) it follows that the characteristic function can also be factored as

$$C(\omega) = \langle \exp(i\omega X_1) \rangle \langle \exp(i\omega X_2) \rangle \cdots \langle \exp(i\omega X_n) \rangle. \tag{1.4.53}$$

If we denote the characteristic function of the variable X_i by $C_i(\omega)$ then eqn(1.4.53) can be written as the product

$$C(\omega) = \prod_{j=1}^{n} C_j(\omega). \tag{1.4.54}$$

The decomposition property for independent random variables is valid for vectors as well as for scalars.

An important consequence of the factorization property in eqn(1.4.54) is that cumulants of independent random variables are additive. On taking the logarithm of the characteristic function as expressed in eqn(1.4.54) and using the definition of the cumulant generating function in eqn(1.4.34) we find that

$$K(\omega) = \sum_{j=1}^{n} K_j(\omega), \tag{1.4.55}$$

where $K_j(\omega) \equiv \ln C_j(\omega)$. Let the mth cumulant of the sum be denoted by $\kappa_m(Y)$, and let the mth cumulant of x_j be denoted by $\kappa_m(X_j)$. Then the decomposition property of eqn(1.4.55) is equivalent to the decomposability of κ_m in the form

$$\kappa_m(Y) = \sum_{j=1}^{n} \kappa_m(X_j). \tag{1.4.56}$$

In particular, since the second cumulant can be identified as a variance, it follows that the variance of a sum of independent random variables is equal to the sum of the variances of the component variables. This property will be used repeatedly in a number of our later calculations.

It is often of interest to calculate cumulants of a linear combination of independent random variables, where the coefficients of the latter variables are known constants. Let such a combination be given by

$$Y = \sum_{j=1}^{n} a_j X_j, \tag{1.4.57}$$

where $\{a_j\}$ are constants and $\{X_j\}$ are independent random variables. The characteristic function corresponding to Y is given by

$$\begin{aligned} C(\omega) &= \prod_{j=1}^{n} \langle \exp(i\omega a_j X_j) \rangle \\ &= \prod_{j=1}^{n} C_j(a_j\omega). \end{aligned} \tag{1.4.58}$$

The cumulant generating function for a term of the sum is

$$K_j(a_j\omega) \equiv \sum_{n=1}^{\infty} \frac{i^n \kappa_n(X_j) a_j^n \omega^n}{n!} = \ln C_j(a_j\omega), \tag{1.4.59}$$

where $\kappa_n(X_j)$ is the cumulant of X_j. It follows that the mth cumulant of Y is given by

$$\begin{aligned} \kappa_m(Y) &= \sum_{j=1}^{n} \kappa_m(a_j X_j) \\ &= \sum_{j=1}^{n} a_j^m \kappa_m(X_j). \end{aligned} \tag{1.4.60}$$

The last line is a consequence of the relation in eqn(1.4.59). This result will be applied in Chapter 4 to the calculation of exact cumulants and moments of the structure factor, which are of importance for generalized intensity statistics.

As a simple application of statistical independence, as discussed in this section, we give a brief derivation of the characteristic function corresponding to the pdf of the normalized structure factor in $P\bar{1}$. A representation of $E(\mathbf{h})$, in this space group, is shown in (1.4.21). If the θs — i.e. the atomic phase factors — are assumed to be independent random variables, the characteristic function of $E(\mathbf{h})$ is given by

$$\langle \exp[i\omega E(\mathbf{h})] \rangle = \prod_{j=1}^{N/2} \langle \exp(i\omega n_j \cos\theta) \rangle = \prod_{j=1}^{N/2} J_0(2\omega n_j), \tag{1.4.61}$$

where the second term is a consequence of the assumed independence of the θs and the third term follows from the explicit representation of the characteristic function in eqn(1.4.38). Further examples of these ideas are given in the following chapter and the principles outlined in this section will be used extensively throughout the monograph.

2

SOME ELEMENTARY PDFS

As has been pointed out in the previous chapter, much of the theory of crystallographic statistics is concerned with the construction of both approximate and exact pdfs for the structure factor. Before we embark on a more systematic exposition of existing approaches to the calculation of these pdfs, it is instructive to present some examples of the use of probability theory in deriving exact, although possibly not easily computable, pdfs for use in intensity statistics. We will reduce these exact results to their limiting forms as the number of atoms in the unit cell becomes large and the scattering factors are not too dissimilar. We include in this chapter a brief discussion of the analogy between problems of crystallographic statistics and some aspects of the theory of random walks.

2.1 The pdf of $|E|$ in space group $P1$

Our first example deals with the calculation of the pdf of the magnitude of the normalized structure factor, $|E|$, in the non-centrosymmetric space group $P1$. The calculation will be made using simplifying assumptions such as the absence of non-crystallographic symmetry and the absence of dispersive scatterers. We also assume, as is done in almost all the parts of this monograph, that the atomic coordinates do not approximate to small fractions or, more precisely, that they are rationally independent as defined in eqn(1.3.19). The normalized structure factor in space group $P1$, under our earlier assumptions, is given by

$$E(\mathbf{h}) = \sum_{j=1}^{N} n_j \exp(i\theta_j), \tag{2.1.1}$$

where $\theta_j = 2\pi\mathbf{h}^T\mathbf{r}_j$, $\mathbf{h}$ is the diffraction vector satisfying the equations of Laue, and $n_j = f_j/(\sum_{k=1}^{N} f_k^2)^{1/2}$, where f_j is the conventional scattering factor of the jth atom. Since the exponentials in (2.1.1) are complex numbers, the normalized structure factor can be regarded as a sum of vectors in the complex plane. It is customary to represent the structure factor diagrammatically (see Fig. 2.1) as a vector polygon, and draw an analogy between this polygon and a two-dimensional random walk, first suggested

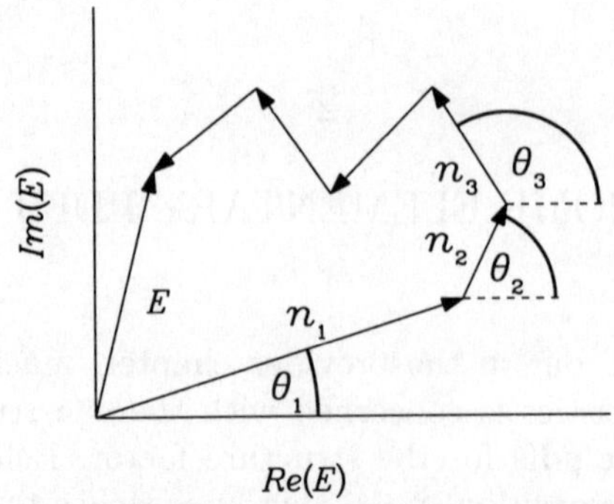

FIG. 2.1. *The vector-polygon representation of the structure factor.* The analogy to the Pearson random walk corresponds to θ being uniformly distributed in $(-\pi, \pi)$.

by Pearson (1905) and introduced into crystallographic literature by Hauptman and Karle (1952) and by Wilson (1952).

Indeed, if we compare the atomic scattering factor to the length of a step and the atomic phase factor θ_j to the angle made by this step with the horizontal axis (see Fig. 2.1), the probability of finding the magnitude of the normalized structure factor between $|E|$ and $|E| + d|E|$ is equal to the probability of finding the random walker at a distance between $|E|$ and $|E| + d|E|$ from the starting point of the walk. Clearly, the probabilistic basis of this comparison is the assumption that the phase factor θ_j is a random variable. The series of randomly oriented steps that comprise the random walk is then equivalent to the series of randomly phased atomic contributions.

In order to calculate the pdf of E, we decompose it into its real and imaginary parts

$$E = A + iB, \tag{2.1.2}$$

and find the equivalent representation in polar coordinates. In space group $P1$ the components of A and B are given by

$$A = \sum_{j=1}^{N} n_j \cos\theta_j \equiv |E| \cos\varphi \tag{2.1.3}$$

and

$$B = \sum_{j=1}^{N} n_j \sin\theta_j \equiv |E| \sin\varphi \tag{2.1.4}$$

where φ is the phase of the structure factor. We wish to find the probability that the magnitude of the normalized structure factor lies between $|E|$ and $|E|+d|E|$. We shall do this by finding the joint pdf of A and B, as given by eqns (2.1.3) and (2.1.4), and integrating out the phase φ from the result. The method of characteristic functions, as developed earlier in section 1.4, will be used in this calculation.

The characteristic function of the (A, B) pair is given by

$$\begin{aligned} C(\omega_1, \omega_2) &= \int_{-\infty}^{\infty} \int_{-\infty}^{\infty} \exp[i(\omega_1 A + \omega_2 B)] p(A, B) dA dB \\ &= \langle \exp[i(\omega_1 A + \omega_2 B)] \rangle. \end{aligned} \tag{2.1.5}$$

The average in the second part of eqn(2.1.5) will be evaluated using the following assumptions:

(i) the atomic contributions to E are statistically independent, and
(ii) the atomic phase angles θ_j in eqns (2.1.3) and (2.1.4) are uniformly distributed in $(-\pi, \pi)$.

If we substitute for A and B their exact representations from eqns (2.1.3) and (2.1.4) we obtain

$$C(\omega_1, \omega_2) = \left\langle \exp\left[i \sum_{j=1}^{N} n_j (\omega_1 \cos\theta_j + \omega_2 \sin\theta_j) \right] \right\rangle \tag{2.1.6}$$

$$= \prod_{j=1}^{N} \langle \exp[i n_j (\omega_1 \cos\theta_j + \omega_2 \sin\theta_j)] \rangle \tag{2.1.7}$$

where assumption (i) has been used. Let us introduce polar coordinates in the transform variables by setting $\omega_1 = \omega \cos\Delta$, $\omega_2 = \omega \sin\Delta$, where $\omega = (\omega_1^2 + \omega_2^2)^{1/2}$ and $\tan\Delta = \omega_2/\omega_1$. We can rewrite eqn(2.1.7) as

$$C(\omega, \Delta) = \prod_{j=1}^{N} \langle \exp[i n_j \omega (\theta_j - \Delta)] \rangle. \tag{2.1.8}$$

The assumption that atomic phase angles are uniformly distributed [(ii) above] permits us to express the average in eqn(2.1.8) by an integral averaged over the variable θ using uniform weighting. We therefore can write

$$C(\omega, \Delta) = \prod_{j=1}^{N} \left\{ \frac{1}{2\pi} \int_{-\pi}^{\pi} \exp[i n_j \omega \cos(\theta - \Delta)] d\theta \right\}. \tag{2.1.9}$$

Since the cosine term in the exponent is periodic we can ignore the Δ in the argument and are left with an integral defining a Bessel function (e.g. Abramowitz and Stegun 1972). Thus $C(\omega_1, \omega_2)$ depends only on the amplitude ω and can be written

$$C(\omega) = \prod_{j=1}^{N} J_0(\omega n_j), \tag{2.1.10}$$

where $J_0(x)$ is the Bessel function of the first kind and of order zero. The joint pdf of A and B is given by the inverse Fourier transform of the characteristic function (see section 1.4) and hence

$$p(A, B) = \int_{-\infty}^{\infty}\int_{-\infty}^{\infty} C(\omega_1, \omega_2) \exp[-i(\omega_1 A + \omega_2 B)] d\omega_1 d\omega_2. \tag{2.1.11}$$

If we use eqns (2.1.3) and (2.1.4), the exponential in eqn(2.1.11) can be represented as

$$\begin{aligned}\exp[-i(\omega_1 A + \omega_2 B)] &= \exp[-i|E|(\omega_1 \cos\varphi + \omega_2 \sin\varphi)] \\ &= \exp[-i|E|\omega \cos(\varphi - \Delta)],\end{aligned} \tag{2.1.12}$$

where an analogous transformation to polar coordinates has been made as in the calculation of the characteristic function. The volume element, $d\omega_1 d\omega_2$, in eqn(2.1.11) now transforms to $\omega d\omega d\Delta$ and eqn(2.1.11) becomes

$$p(|E|\cos\varphi, |E|\sin\varphi) = \frac{1}{4\pi^2}\int_0^{\infty} \omega C(\omega) d\omega \int_{-\pi}^{\pi} \exp[-i|E|\omega\cos(\varphi - \Delta)] d\Delta. \tag{2.1.13}$$

On integrating out the angular variable Δ (φ is fixed for this pair of A and B) we obtain

$$p(|E|\cos\varphi, |E|\sin\varphi) = \frac{1}{2\pi}\int_0^{\infty} \omega \left[\prod_{j=1}^{N} J_0(\omega n_j)\right] J_0(\omega|E|) d\omega. \tag{2.1.14}$$

We denote the joint pdf of the variables $|E|$ and φ by $g(|E|, \varphi)$. It is readily seen that $g(|E|, \varphi) = |E| p(|E|\cos\varphi, |E|\sin\varphi)$, and the required pdf of $|E|$, denoted by $p(|E|)$, is obtained by integrating out the phase from $g(|E|, \varphi)$, that is to say

$$p(|E|) = \int_{-\pi}^{\pi} g(|E|, \varphi) d\varphi. \tag{2.1.15}$$

Since the integrand in eqn(2.1.14) is independent of φ the integration over angle yields another factor of 2π. We finally obtain an expression for the pdf of $|E|$ in the form

$$p(|E|) = \left\{ |E| \int_0^\infty \omega \left[\prod_{j=1}^{N} J_0(\omega n_j) \right] J_0(\omega|E|) d\omega \right\}. \quad (2.1.16)$$

Equation (2.1.16) is analogous to Kluyver's (1906) solution of Pearson's (1905) problem of the two-dimensional random walk, and was introduced into the crystallographic literature by Hauptman and Karle (1952), who used the cumulative distribution rather than the pdf. However, a numerical calculation of the value $p(|E|)$ based on the integral representation presents considerable difficulties because of the oscillatory integrand. Analogous expressions for the pdf of $|E|$ can, of course, be found for other non-centrosymmetric space groups when appropriate functional forms are used for the trigonometric structure factor in calculating the characteristic function.

In the context of crystallographic statistics it is useful to investigate the limiting form of eqn(2.1.16) in the limit of a large number of identical atoms in the unit cell of $P1$. If all the atoms are the same the normalized scattering factor reduces to $N^{-1/2}$ and eqn(2.1.16) can be written in the form

$$p(|E|) = |E| \int_0^\infty \omega J_0^N(\omega N^{-1/2}) J_0(\omega|E|) d\omega. \quad (2.1.17)$$

The Nth power of the Bessel function can be written as

$$\begin{aligned} J_0^N(\omega N^{-1/2}) &= \exp\left[N \ln J_0(\omega N^{-1/2})\right] \\ &= \exp\left[N \ln\left(1 - \frac{\omega^2}{4N} + \frac{\omega^4}{64N^2} - \cdots\right)\right]. \end{aligned} \quad (2.1.18)$$

When N is large and the logarithm is expanded in a Taylor series the first term is approximately proportional to ω^2/N and the second to ω^4/N^2. Consequently, when ω is fixed and N is allowed to increase we can write

$$J_0^N(\omega N^{-1/2}) \sim \exp\left[-\frac{\omega^2}{4}\right]. \quad (2.1.19)$$

The expression for $p(|E|)$ is then

$$p(|E|) \simeq |E| \int_0^\infty \omega \exp\left[-\frac{\omega^2}{4}\right] J_0(\omega|E|) d\omega = 2|E| \exp(-|E|^2) \quad (2.1.20)$$

[Gradshteyn and Ryzhik 1980, eqn(6.631[4]); see also Appendix A, eqn(A.0.1)]. The expression in the right-hand side of (2.1.20) is just the ideal acentric pdf derived by Wilson (1949) on the basis of the central limit theorem or Gaussian approximation (see Chapter 3). Later we shall also discuss other ways to improve the accuracy of this type of approximation as well as a more convenient starting point from which to obtain exact representations of the pdf. We point out, however, that in the present case corrections to the Gaussian result can be found by retaining higher-order terms in the power series expansions of the Bessel functions in the integral in eqn(2.1.17). A somewhat more systematic approach to deriving corrections to the Gaussian approximation to pdfs will be discussed in Chapter 4.

2.2 The pdf of $|E|$ in space group $P\bar{1}$

Our second example deals with the derivation of the pdf of $|E|$ for the centrosymmetric space group $P\bar{1}$. We shall make the same assumptions as in the derivation of eqn(2.1.16). The normalized structure factor for this space group is given by

$$E(\mathbf{h}) = 2\sum_{j=1}^{N/2} n_j \cos\theta_j. \tag{2.2.21}$$

The above equation is conveniently pictured as a sum of pairs of atomic contribution, each contribution corresponding to atoms at $\mathbf{r}_j$ and $-\mathbf{r}_j$. Since $E(\mathbf{h})$ now consists only of a real part, the analogous random walk is one-dimensional consisting of steps randomly directed in the positive or negative directions along the real axis.

The solution of the problem of characterizing random properties of $E(\mathbf{h})$ clearly starts from the derivation of a univariate pdf, which we denote by $p(E)$. In order to emphasize the fact that this is a one-dimensional problem we use the single transform variable ω_1 in place of the vector pair (ω_1, ω_2). The characteristic function corresponding to the (univariate) pdf of $E(\mathbf{h})$ is now given by

$$C(\omega_1) = \int_{-\infty}^{\infty} p(E)\exp(i\omega_1 E)dE \equiv \langle \exp(i\omega_1 E)\rangle, \tag{2.2.22}$$

which may be evaluated by directly substituting the expression in eqn (2.2.21) into eqn(2.2.22). We have

$$C(\omega_1) = \left\langle \exp\left(2i\omega_1 \sum_{j=1}^{N/2} n_j \cos\theta_j\right)\right\rangle \tag{2.2.23}$$

$$= \left\langle \prod_{j=1}^{N/2} \exp(2i\omega_1 n_j \cos\theta_j)\right\rangle \tag{2.2.24}$$

$$= \prod_{j=1}^{N/2} \langle \exp(2i\omega_1 n_j \cos\theta) \rangle \tag{2.2.25}$$

$$= \prod_{j=1}^{N/2} \left\{ \frac{1}{2\pi} \int_{-\pi}^{\pi} \exp(2i\omega_1 n_j \cos\theta) d\theta \right\} \tag{2.2.26}$$

$$= \prod_{j=1}^{N/2} J_0(2\omega_1 n_j). \tag{2.2.27}$$

Equation (2.2.25) follows from eqn(2.2.24) by the assumption of independence, while eqn(2.2.26) is obtained from eqn(2.2.25) by invoking the uniform distribution of the atomic phase angles.

Since the pdf is an inverse Fourier transform of the characteristic function

$$p(E) = \frac{1}{2\pi} \int_{-\infty}^{\infty} C(\omega_1) \exp(-i\omega_1 E) d\omega_1$$

[see eqn(1.4.29)] we obtain, using eqns (2.2.25)–(2.2.27)

$$p(E) = \frac{1}{2\pi} \int_{-\infty}^{\infty} \left[\prod_{j=1}^{N/2} J_0(2\omega_1 n_j) \right] \exp(-i\omega_1 E) d\omega_1. \tag{2.2.28}$$

As in the previous example, this exact result is analogous to the solution by Kluyver (1906) of the problem (Pearson 1905) of a one-dimensional random walk. The crystallographic analogy was first discussed by Hauptman and Karle (1952). However, we again deal with a highly oscillatory integrand, suggesting the use of an approximation similar to that used in our analysis of the pdf for the non-centrosymmetric case. We point out that Kluyver's method of obtaining the solution is appreciably more complicated than the use of characteristic functions in the present exposition.

Let us investigate the behaviour of eqn(2.2.28) in the simple case in which the asymmetric unit of $P\bar{1}$ consists of a large number of identical atoms. As in the previous example, when all of the atoms are identical the characteristic function becomes

$$C(\omega_1) = J_0^{N/2}(2\omega_1 N^{-1/2}). \tag{2.2.29}$$

When ω_1 is fixed and N is large we may use the same approximation as in in eqn(2.1.19) to conclude that

$$C(\omega_1) \sim \exp\left(-\frac{\omega_1^2}{2}\right). \tag{2.2.30}$$

This allows us to approximate the (relevant) real part of the exact pdf of E by

$$p(E) = \frac{1}{2\pi} \int_{-\infty}^{\infty} \exp\left(-\frac{\omega_1^2}{2}\right) \cos(\omega_1 E) d\omega_1 = \frac{1}{\sqrt{2\pi}} \exp\left(-\frac{E^2}{2}\right). \quad (2.2.31)$$

[Gradshteyn and Ryzhik (1980), eqn(3.896[4]); see also Appendix A, eqn(A.0.2)]. In centrosymmetric space groups, with the space group origin at a centre of symmetry, we expect to have $p(E) = p(-E)$, which implies that

$$p(|E|) = \sqrt{\frac{2}{\pi}} \exp\left(-\frac{|E|^2}{2}\right). \quad (2.2.32)$$

This is just the ideal pdf for a centric distribution, derived by Wilson (1949) on the basis of the central limit theorem.

In this chapter we have shown how the formalism of probability theory can be applied to the construction of exact pdfs of $|E|$ and how, under appropriate circumstances, these exact representations can be reduced to limiting forms. These are of considerable importance in crystallographic practice in the resolution of space-group ambiguities under fairly common conditions. When, however, the assumption of a 'large number of nearly-equal atoms' is not valid, either higher approximations to the limiting pdfs or more convenient exact representations of the pdfs are necessary. Chapters 4 and 5 are devoted to these generalized aspects of structure-factor statistics.

3

IDEAL DISTRIBUTIONS

3.1 The central limit theorem

In our explicit calculation of the asymptotic form of the pdfs for the magnitude of the structure factor in space groups $P1$ and $P\bar{1}$ we showed that the characteristic function for the equal-atom case approaches a pure or modified Gaussian form when the number of atoms in the unit cell is large. In this section we show that the Gaussian limiting form is to be expected under more general circumstances than the equal-atom case for the two space groups analysed in Chapter 2. Indeed, the Gaussian limit is nearly always the starting point for investigations of both intensity statistics and direct methods in the crystallographic literature. In the mathematical literature the circle of ideas related to this Gaussian behaviour is known as the *central limit theorem* (e.g. Cramér 1946; Feller 1971). This theorem first appeared with specific reference to crystallographic applications in a paper by Wilson (1949). The results of this investigation are generally referred to as the 'Wilson statistics', and have had a great influence on the further development of crystallographic statistics including direct methods. Since the central limit theorem plays such an important role in crystallography it is useful to discuss some of the more elementary ideas behind this much used tool.

To do so, we start by considering the simplest possible situation in which one is interested in finding an approximate form for the pdf of a sum of N independent, identically distributed, random variables. An example to be treated in following sections is that of finding the pdf of the structure factor for a crystal, the asymmetric unit of which consists of N identical atoms with rationally independent coordinates.

To begin with we consider, without reference to any specific crystallographic application, properties of the sum, Y, of N scalar independent and identically distributed random variables, $X_1, X_2, \ldots, X_N$:

$$Y = X_1 + X_2 + \cdots + X_N. \tag{3.1.1}$$

Let $p(x)$ be the pdf of any one of the Xs and let $C(\omega)$ be the corresponding characteristic function or Fourier transform of $p(x)$ [see eqn(1.4.28)]. A knowledge of the characteristic function of Y, call it $C_Y(\omega)$, enables one to calculate the pdf of Y by using the inversion formula given in eqn(1.4.29). Since $p(x)$ is a known function, $C(\omega)$ can be found from it. Let us calculate the composite characteristic function, $C_Y(\omega)$, in terms of the characteristic function for a single x, $C(\omega)$. We have

$$\begin{aligned} C_Y(\omega) &= \langle \exp[i\omega(X_1 + X_2 + \cdots + X_N)] \rangle \\ &= \int_{-\infty}^{\infty} \cdots \int_{-\infty}^{\infty} \exp[i\omega(x_1 + x_2 + \cdots + x_N)] \\ &\quad \times p(x_1)p(x_2)\ldots p(x_N)dx_1 dx_2 \ldots dx_N \\ &= \prod_{j=1}^{N} \left[\int_{-\infty}^{\infty} \exp(i\omega x_j)p(x_j)dx_j \right] \\ &= C^N(\omega), \end{aligned} \tag{3.1.2}$$

where the factorization property is a consequence of the assumed independence of the Xs. The identity of the integrals in the fourth line of (3.1.2) follows from the assumed identical distribution of the variables. Thus, the pdf of Y, $q(Y)$, can be formally represented as

$$q(Y) = \frac{1}{2\pi} \int_{-\infty}^{\infty} C^N(\omega) \exp(-i\omega Y) d\omega. \tag{3.1.3}$$

Since, as we have already pointed out, $|C(\omega)| \leq 1$ for $\omega \neq 0$ and $C(0) = 1$, the value of the integral, as N becomes increasingly large, will be dominated by the contributions from values of ω that are close to 0. Thus, one strategy for finding an approximate form of the function, $q(Y)$, will be to approximate the function $C^N(\omega)$ carefully in the neighbourhood of $\omega = 0$, and assume that the same functional form can be used for values of ω that are not necessarily small. Another technique for finding an approximation to $q(Y)$, the saddlepoint approximation or method of steepest descents, will be described in section 1 of Chapter 4.

In order to carry out this program for cases relevant to crystallographic applications let us note that all of the $p(x)$ that one meets are symmetric around $x = 0$, which means that they have the property

$$\langle X^n \rangle \equiv \int_{-\infty}^{\infty} x^n p(x) dx = 0, \tag{3.1.4}$$

for all odd n. For example, in our discussion of space groups $P1$ and $P\bar{1}$, the pdf associated with any one of the random variables θ appearing in (2.1.1) and (2.2.21) is $p(\theta) = (2\pi)^{-1}$ for $-\pi \leq \theta < \pi$. Since the odd moments of $p(x)$ are all equal to zero one can expand $C(\omega)$ as [cf. (1.4.32)]:

$$C(\omega) = \sum_{n=0}^{\infty} (-1)^n \frac{\omega^{2n}}{(2n)!} \langle X^{2n} \rangle. \tag{3.1.5}$$

To find the lowest-order approximation to $C^N(\omega)$ for small ω we express this function as

$$C^N(\omega) = \exp[N \ln C(\omega)] = \exp\left[N \ln\left(1 - \frac{\omega^2}{2!}\langle X^2\rangle + \frac{\omega^4}{4!}\langle X^4\rangle - \ldots\right)\right]. \tag{3.1.6}$$

Since only the small ω behaviour of $C^N(\omega)$ is of interest we expand the logarithm around $\omega = 0$, while retaining only the lowest-order term. This gives rise to an approximation to $C^N(\omega)$ in the form of a Gaussian

$$C^N(\omega) \sim \exp\left(-N\frac{\langle X^2\rangle}{2}\omega^2\right). \tag{3.1.7}$$

When this function is inserted into (3.1.3) one finds

$$\begin{aligned} q(Y) &\sim \frac{1}{2\pi}\int_{-\infty}^{\infty} \exp\left(-N\frac{\langle X^2\rangle}{2}\omega^2 - i\omega Y\right) d\omega \\ &= \frac{1}{\sqrt{2\pi N\langle X^2\rangle}} \exp\left(-\frac{Y^2}{2N\langle X^2\rangle}\right). \end{aligned} \tag{3.1.8}$$

This is the basic result of the simplest version of the central limit theorem. The significant feature of the result is that the form of $q(Y)$ is independent of the underlying pdfs of the Xs, depending only on a single parameter calculated from that pdf, namely the variance, $\langle X^2\rangle$. The reader will undoubtedly notice that the calculation given in eqn(3.1.8) closely follows what we have done in finding an approximate form for $p(|E|)$ in eqns (2.1.17)–(2.1.20). This observation implies that the Gaussian form of the pdf arising from the central limit theorem can be applied to a much broader class of problems than that of pdfs of the structure-factor magnitudes in the space groups $P1$ and $P\bar{1}$ as dealt with in Chapter 2.

Three parenthetic remarks may be made about the Gaussian approximation in eqn(3.1.8). The first is that it is convenient in comparing several pdfs to fix the variance for all of them to have the same value. Conventionally, the variance is set equal to 1 which amounts to calculating the pdf of the *normalized* random variable, ζ, defined by

$$\zeta = \frac{Y}{\sqrt{N\langle X^2\rangle}} = \frac{X_1 + X_2 + \cdots + X_N}{\sqrt{N\langle X^2\rangle}}, \tag{3.1.9}$$

whose variance, $\langle \zeta^2\rangle$, is readily seen to be equal to 1. The central limit theorem is equivalent to the statement that in the limit $N \to \infty$ the pdf of ζ, $q(\zeta)$, approaches the Gaussian form

$$q(\zeta) = \frac{1}{\sqrt{2\pi}} \exp\left(-\frac{\zeta^2}{2}\right). \tag{3.1.10}$$

This result depends on the finiteness of $\langle X^2 \rangle$, which will always be the case in crystallographic applications.

Our second remark is that even though the Xs are bounded, in the sense that each random variable can only take values within some interval, say $(-S, S)$ so that Y can only take on values in $(-NS, NS)$, the Gaussian pdf in eqns (3.1.8) or (3.1.10) allows Y to take on any value in the infinite interval $(-\infty, \infty)$. In all structure-factor calculations the xs are necessarily bounded because they are linear combinations of products of trigonometric functions and scattering factors. It follows from these observations that the Gaussian approximation is not expected to be accurate in the tails of the pdf but may be expected to be accurate in the vicinity of the peak provided, of course, that the assumptions that the random variables are independent and identically distributed are valid.

Thirdly, one can easily calculate corrections to the lowest-order Gaussian approximation in eqn(3.1.8) by taking account of successive terms in eqn(3.1.6). For example, if we expand the exponential in eqn(3.1.6) keeping terms of up to order ω^4 we have

$$\begin{aligned} C^N(\omega) &\sim \exp\left[-N\frac{\langle X^2\rangle}{2}\omega^2 + \frac{N}{24}\{\langle X^4\rangle - 3\langle X^2\rangle^2\}\omega^4\right] \\ &\sim \exp\left(-N\frac{\langle X^2\rangle}{2}\omega^2\right)\left[1 + \frac{N}{24}\{\langle X^4\rangle - 3\langle X^2\rangle^2\}\omega^4\right]. \end{aligned} \tag{3.1.11}$$

On inserting this expression into eqn(3.1.3) and evaluating the integral, we find, for the first two terms in the expansion of $q(Y)$,

$$\begin{aligned} q(Y) \sim \frac{1}{\sqrt{2\pi N\langle X^2\rangle}} \exp\left(-\frac{Y^2}{2N\langle X^2\rangle}\right)\Bigg[1 + \frac{\langle X^4\rangle - 3\langle X^2\rangle^2}{24N\langle X^2\rangle^2} \\ \times \left(\frac{Y^4}{N^2\langle X^2\rangle^2} - 6\frac{Y^2}{N\langle X^2\rangle} + 3\right) + \cdots\Bigg]. \end{aligned} \tag{3.1.12}$$

A more systematic way of finding higher-order approximations to the exact $q(Y)$ in terms of moments will be developed when we discuss the Edgeworth expansions and related expansions in the next chapter. We shall also see that both the moments of orders higher than the second of structure-factor components, as well as the characteristic function, depend on the degree of atomic heterogeneity and space-group symmetry. However, as long as the contributions from different atoms can be treated as independent and identically distributed random variables, this dependence on space-group symmetry has only minor consequences and the central limit theorem approach can be invoked to furnish a reliable approximation to the pdf of the sum.

Both the appeal of the Gaussian approximation and the way in which it can break down in crystallographic applications are conveniently discussed together. The reason that the central limit theorem is so widely applied is that it requires only a minimal knowledge of properties of the underlying pdfs. Another useful feature of the Gaussian approximation is that when it is valid one can often combine pdfs in a simple fashion allowing one to take atomic heterogeneity into account, provided that the heterogeneity is not too great. To illustrate, suppose we have a unit cell in space group $P\bar{1}$ containing only two types of atoms, M with a normalized scattering factor equal to n_M, and N with a normalized scattering factor equal to n_N, and assume that $Z_M > Z_N$, where Z is an atomic number. The components of the structure factor are explicitly

$$\begin{aligned} A &= 2n_M \sum_{i=1}^{M/2} \cos\theta_i + 2n_N \sum_{j=1}^{N/2} \cos\phi_j \\ B &= 0. \end{aligned} \tag{3.1.13}$$

Assume that both M and N are much greater than 1. Under these circumstances one would expect the pdf of the structure factor from either group of atoms in the absence of the other to be approximately Gaussian. The characteristic function for this example can be found in closed form as

$$C(\omega_1) = \langle \exp(i\omega_1 A)\rangle = J_0^{M/2}(2n_M\omega_1)J_0^{N/2}(2n_N\omega_1). \tag{3.1.14}$$

In the limit of large M and N we can expand each of the Bessel functions in the neighbourhood of $\omega_1 = 0$. Consider the component determined by the group of M atoms. This is

$$\begin{aligned} J_0^{M/2}(2n_M\omega_1) &= \exp\left[\frac{M}{2}\ln J_0(2n_M\omega_1)\right] \\ &\sim \exp\left[\frac{M}{2}\ln\left(1 - n_M^2\omega_1^2\right)\right] \\ &\sim \exp\left[-\frac{Mn_M^2\omega_1^2}{2}\right]. \end{aligned} \tag{3.1.15}$$

This approximation, together with eqn(3.1.14) implies that

$$C(\omega_1) \sim \exp\left[-(Mn_M^2 + Nn_N^2)\frac{\omega_1^2}{2}\right], \tag{3.1.16}$$

so that even with this atomic heterogeneity one still retains the approximation by a Gaussian form for the pdf. If, however, the number of outstandingly heavy atoms is small (let us say, equal to 1) then, even though the total number of atoms in the unit cell might be large, the Gaussian pdf can be a qualitatively poor approximation to the overall pdf, depending on the particular set of parameters. We will return to a discussion of the general problem raised by this scenario as well as ways to tackle its resolution in the next chapter.

The one-dimensional version of the central limit theorem is readily extended to higher dimensions, but the results may look somewhat more complicated when correlations are taken into account.

Let us derive the simplest form of the central limit theorem in two dimensions, which results from the assumption that the second-order correlations of the random variables are equal to zero. In this case it will be seen that the approximate pdf is factorizable in the sense of eqn(1.4.49). We will suppose that there are two sums, Y and Y', defined by

$$Y = X_1 + X_2 + \cdots + X_N, \; Y' = X_1' + X_2' + \cdots + X_N' \tag{3.1.17}$$

where, without losing generality, we may assume that

$$\langle X_j \rangle = \langle X_j' \rangle = 0, \; j = 1, 2, \ldots, N. \tag{3.1.18}$$

Just as is true in one dimension the derivation of the central limit theorem is based on properties of the characteristic function in the neighbourhood of the origin of transform parameter space. We will define the two-dimensional characteristic function, $C(\boldsymbol{\omega}) = C(\omega_1, \omega_2)$, by

$$\begin{aligned} C(\omega_1, \omega)_2) &= \langle \exp[i(\omega_1 Y + \omega_2 Y')] \rangle \\ &= \left\langle \exp\left[i \left(\omega_1 \sum_{j=1}^{N} X_j + \omega_2 \sum_{j=1}^{N} X_j' \right) \right] \right\rangle \\ &= \prod_{j=1}^{N} \langle \exp[i(\omega_1 X_j + \omega_2 X_j')] \rangle \\ &= \langle \exp[i(\omega_1 X + \omega_2 X')] \rangle^N. \end{aligned} \tag{3.1.19}$$

Notice that no further factorization of the characteristic function is possible without the additional assumption that X and X' are independent of one another. For the sake of producing the simplest possible version of the two-dimensional central-limit theorem we will assume that the correlations are equal to zero, later generalizing the result by including effects of second-order correlations.

The strategy for deriving the asymptotic joint pdf of Y and Y' is to expand the function $\langle \exp[i(\omega_1 X + \omega_2 X')] \rangle$ around $\omega_1 = \omega_2 = 0$ and exponentiate the result. In this way we find

$$\langle \exp[i(\omega_1 X + \omega_2 X')] \rangle^N = \exp\left\{ N \ln \langle \exp[i(\omega_1 X + \omega_2 X')] \rangle \right\}$$

$$\sim \exp\left\{N \ln\left(1 - \frac{\langle(\omega_1 X + \omega_2 X')\rangle^2}{2}\right)\right\}$$
$$\sim \exp\left[-\frac{N(\omega_1\sigma^2 + \omega_2\sigma'^2)}{2}\right]. \quad (3.1.20)$$

This allows us to write, for the central limit approximation to the joint pdf $q(Y, Y')$:

$$q(Y,Y') \sim \frac{1}{(2\pi)^2}\int_{-\infty}^{\infty}\int_{-\infty}^{\infty} \exp\left[-\frac{N}{2}(\omega_1^2\sigma^2 + \omega_2^2\sigma'^2)\right] \times \exp[-i(\omega_1 Y + \omega_2 Y')]d\omega_1 d\omega_2$$
$$= \frac{1}{2\pi N\sigma\sigma'}\exp\left[-\frac{1}{2N}\left(\frac{Y^2}{\sigma^2} + \frac{Y'^2}{\sigma'^2}\right)\right]. \quad (3.1.21)$$

We see that even though the random variables Y and Y' might be dependent the pdf is nevertheless factorizable in the Gaussian approximation provided that the second-order correlation functions, $\langle X_j X_j' \rangle$, are equal to zero.

One further result derivable from eqn(3.1.21) in the Gaussian approximation is the pdf of the magnitude

$$\rho = \sqrt{Y^2 + Y'^2}. \quad (3.1.22)$$

To calculate the required pdf we transform to polar coordinates using

$$Y = \rho\cos\beta \quad (3.1.23)$$
$$Y' = \rho\sin\beta \quad (3.1.24)$$

and integrate out the angular variable β. To carry out this program write

$$q(y, y')dydy' = q(\rho\cos\beta, \rho\sin\beta)\rho d\rho d\beta \quad (3.1.25)$$

so that the normalized pdf of the variable ρ, denoted by $h(\rho)$, is

$$h(\rho) = \rho\int_{-\pi}^{\pi} q(\rho\cos\beta, \rho\sin\beta)d\beta \quad (3.1.26)$$
$$= \frac{\rho}{2\pi N\sigma\sigma'}\int_{-\pi}^{\pi}\exp\left[-\frac{\rho^2}{2N}\left(\frac{\cos^2\beta}{\sigma^2} + \frac{\sin^2\beta}{\sigma'^2}\right)\right]d\beta. \quad (3.1.27)$$

On substituting the trigonometric identities $\cos^2\beta = (1 + \cos 2\beta)/2$, and $\sin^2\beta = (1 - \cos 2\beta)/2$ into this integral we transform it into

$$h(\rho) = \frac{\rho}{2\pi N\sigma\sigma'}\exp\left[-\frac{\rho^2}{4N}\left(\frac{1}{\sigma^2} + \frac{1}{\sigma'^2}\right)\right]$$

$$\times \int_{-\pi}^{\pi} \exp\left[\frac{\rho^2}{4N}\left(\frac{1}{\sigma^2} - \frac{1}{\sigma'^2}\right)\cos 2\beta\right] d\beta. \tag{3.1.28}$$

But we may express the integral in terms of a Bessel function of imaginary argument of the second kind, $I_0(a)$, by making use of the definite integral (Gradshteyn and Ryzhik 1980; 8.431-5)

$$\int_{-\pi}^{\pi} \exp[a\cos\beta]d\beta = 2\pi I_0(a). \tag{3.1.29}$$

In this way we find, as the central limit theorem approximation to the pdf of the amplitude ρ,

$$h(\rho) = \frac{\rho}{2N\sigma\sigma'} \exp\left[-\frac{\rho^2}{4N}\left(\frac{1}{\sigma^2} + \frac{1}{\sigma'^2}\right)\right] I_0\left[\left(\frac{1}{\sigma^2} - \frac{1}{\sigma'^2}\right)\frac{\rho^2}{4N}\right] \tag{3.1.30}$$

When $\sigma = \sigma'$ the Bessel function is equal to unity and $h(\rho)$ reduces to the somewhat simpler form

$$h(\rho) = \frac{\rho}{2N\sigma^2} \exp\left(-\frac{\rho^2}{2N\sigma^2}\right). \tag{3.1.31}$$

As shown in the following paragraphs it is possible to extend our analysis of the Gaussian approximation to the pdf for correlated multivariate random variables, but the basic ideas remain unchanged.

The same general technique which we have used to derive a Gaussian approximation for the pdf for a sum of N scalar random variables can be extended to derive a Gaussian approximation for the sum of N vectors in the limit $N \to \infty$. A new feature in such pdfs is the possibility of correlations. We will derive the form of the Gaussian pdf in two dimensions and later indicate the form of the d-dimensional generalization. Such a generalization is quite important in the implementation of direct methods, which generally require a consideration of correlated variables. There are applications of the same ideas in intensity statistics.

Denote a random two-dimensional vector by $\mathbf{r} = (X, X')$, where the pdf associated with the displacement in a single step will be denoted by $p(x, x')$. This density can be used to generate moments, including those of mixed order. The rsth moment will be written as $\langle X^r(X')^s\rangle$, which is defined by

$$\langle X^r(X')^s\rangle = \int_{-\infty}^{\infty}\int_{-\infty}^{\infty} x^r(x')^s p(x, x')dxdx', \quad r, s = 0, 1, 2, \ldots \tag{3.1.32}$$

We consider the problem of finding an approximation to the pdf of the components of a sum $\mathbf{S}(y, y') = (S_1, S_2)$, where

$$\mathbf{S} = \mathbf{r}_1 + \mathbf{r}_2 + \cdots + \mathbf{r}_n \tag{3.1.33}$$
$$y = x_1 + x_2 + \cdots + x_n \tag{3.1.34}$$
$$y' = x_1{}' + x_2{}' + \cdots + x_n{}'. \tag{3.1.35}$$

(In the context of intensity statistics, the components X and X' may be thought of as the real and imaginary parts of an atomic contribution to the normalized structure factor $E = A + iB$, while y_1 and y_2 correspond to A and B respectively.) For simplicity we only treat the case in which the **r**s are independent, identically distributed random variables. The results can be extended to non-identically distributed variables in a fairly obvious way. This extension is important in crystallographic applications because it pertains to the frequently-occurring situation in which there are dissimilar contributions from each atomic species.

The strategy of deriving the central limit theorem in two dimensions including the effects of second-order correlations is a straightforward extension of the analysis given when these correlations are equal to 0. It is based on the characteristic function as defined in the first line of eqn(3.1.19). Just as in the case of random variables that are scalar, the derivation is based on the expansion of $C(\boldsymbol{\omega})$ around $\boldsymbol{\omega} = \mathbf{0}$ retaining only the lowest-order terms. Let $p(y, y')$ be the joint pdf of two sums, y and y'. We can again assume that the variables have been defined so that Y and Y' have a mean equal to zero, which is to say that $\langle Y \rangle = \langle Y' \rangle = 0$. It will usually be the case in single-structure-factor statistics that because of symmetry the first moments of the relevant random variables are identically equal to zero. Thus, more generally, while the odd moments of the magnitude of the structure factor, $|E|$, always exist, those of the real or imaginary parts of E, or the odd moments of (real) centrosymmetric structure factors, E, usually vanish. This is true for all three-dimensional space groups. However, there are some non-vanishing odd moments that arise from two-dimensional *plane* groups describing projections of trigonal and hexagonal structures (Foster and Hargreaves 1963; Shmueli and Kaldor 1983). We will use the assumption that the first moments vanish in our present analysis and remarking that even when this assumption is not satisfied only trivial modifications of the theory need be made to find the final form of the central limit approximation to $p(\mathbf{S})$. If we expand $C(\boldsymbol{\omega})$ to second order in the ωs we find

$$C(\boldsymbol{\omega}) \simeq 1 - \frac{1}{2}(\langle X^2 \rangle \omega_1^2 + 2\langle XX' \rangle \omega_1 \omega_2 + \langle (X')^2 \rangle \omega_2^2), \tag{3.1.36}$$

which not only contains contributions from the second moments but also from the correlation function $\langle XX' \rangle$ as well. The formula for $p(\mathbf{S})$ is then given by

$$
\begin{aligned}
p(\mathbf{S}) &= \frac{1}{4\pi^2}\int_{-\infty}^{\infty}\int_{-\infty}^{\infty} C^N(\boldsymbol{\omega})\exp(-i\mathbf{S}^T\boldsymbol{\omega})d^2\boldsymbol{\omega} \\
&= \frac{1}{4\pi^2}\int_{-\infty}^{\infty}\int_{-\infty}^{\infty} \exp[N\ln C(\boldsymbol{\omega}) - i\mathbf{S}^T\boldsymbol{\omega}]d^2\boldsymbol{\omega} \\
&\sim \frac{1}{4\pi^2}\int_{-\infty}^{\infty}\int_{-\infty}^{\infty} \exp\left[-\frac{N}{2}\left(\langle X^2\rangle\omega_1^2 + 2\langle XX'\rangle\omega_1\omega_2 + \langle (X')^2\rangle\omega_2^2\right)\right] \\
&\quad \times \exp[-i\mathbf{S}^T\boldsymbol{\omega}]d^2\boldsymbol{\omega}. \qquad (3.1.37)
\end{aligned}
$$

Notice that the quadratic form that appears in this exponent can also be expressed as $\boldsymbol{\omega}^T\mathbf{M}\boldsymbol{\omega}$, in which $\mathbf{M}$, referred to as the variance–covariance matrix, has the form

$$
\mathbf{M} = \begin{pmatrix} \langle X^2\rangle & \langle XX'\rangle \\ \langle XX'\rangle & \langle (X')^2\rangle \end{pmatrix}. \qquad (3.1.38)
$$

An important observation is that since $\mathbf{M}$ is symmetric it can always be diagonalized by a similarity transformation (e.g. Strang 1988). In order to evaluate the integral in eqn(3.1.37) we need the fact that the determinant of $\mathbf{M}$, a quantity denoted by $|\mathbf{M}|$ which is equal to $\langle X^2\rangle\langle (X')^2\rangle - \langle XX'\rangle^2$, is not equal to zero and is in fact strictly positive for pdfs which are not delta functions. This follows from the fact that for any value of α the quantity $\langle (X\alpha - X')^2\rangle$ is non-negative except when $p(x, x')$ consists of a single delta function.

The evaluation of the integral in eqn(3.1.37) is most readily carried out by reducing the quadratic form to a sum of two squares by eliminating the mixed term. To do this we transform to a new set of coordinates (ξ_1, ξ_2) by

$$
\omega_1 = \xi_1 - \frac{\langle XX'\rangle}{\langle X^2\rangle}\xi_2, \; \omega_2 = \xi_2, \qquad (3.1.39)
$$

which transforms the quadratic form to

$$
\boldsymbol{\omega}^T\mathbf{M}\boldsymbol{\omega} = \xi_1^2 + \left(\langle (X')^2\rangle - \frac{\langle XX'\rangle^2}{\langle X^2\rangle}\right)\xi_2^2. \qquad (3.1.40)
$$

In this way the double integral in eqn(3.1.37) is simplified by factoring it into a product of two easily evaluated single integrals with respect to the ξs. The form of the final result is simplified through the introduction of the normalized correlation function ρ as

$$
\rho = \frac{\langle XX'\rangle}{\sqrt{\langle X^2\rangle\langle (X')^2\rangle}}, \qquad (3.1.41)
$$

which satisfies the inequality: $\rho^2 \leq 1$. If we eliminate $\langle XX'\rangle$ in terms of ρ and evaluate the resulting integral we find

$$p(\mathbf{S}) \sim \frac{1}{2\pi N[\langle X^2\rangle\langle (X')^2\rangle(1-\rho^2)]^{1/2}}$$
$$\times \exp\left\{-\frac{1}{2N(1-\rho^2)}\left[\frac{S_1^2}{\langle X^2\rangle} - \frac{2\rho S_1 S_2}{(\langle X^2\rangle\langle (X')^2\rangle)^{1/2}} + \frac{S_2^2}{\langle (X')^2\rangle}\right]\right\}. \tag{3.1.42}$$

In a similar fashion, in d dimensions the characteristic function will be an exponential of a quadratic form. If the components of a single vector step in eqn(3.1.37) are $\mathbf{r}(X_1, X_2, \ldots, X_d)$ then the quadratic form can be expressed as $\boldsymbol{\omega}^T\mathbf{M}\boldsymbol{\omega}$, where the $d \times d$ variance–covariance matrix is composed of the elements

$$(\mathbf{M})_{ij} = \langle X_i X_j\rangle. \tag{3.1.43}$$

The idea behind a derivation of the general case has been suggested in our evaluation of the inversion integral for $d = 2$. The key step is to factor the d-fold integral into a product of d single integrals by making a linear transformation to diagonalize the matrix $\mathbf{M}$. When this is done and the result is substituted into the Fourier inversion integral the expression analogous to eqn(3.1.42) is found to be expressible in terms of $\mathbf{M}$ as

$$p(\mathbf{S}) \sim \frac{1}{(2\pi N)^{d/2}|\mathbf{M}|^{1/2}} \exp\left[-\frac{1}{2N}(\mathbf{S}^T\mathbf{M}^{-1}\mathbf{S})\right], \tag{3.1.44}$$

which is a d-dimensional Gaussian. When $d = 2$ this reduces to the result given in eqn(3.1.42).

Two final points deserve mention. When the random variables in the sum in eqn(3.1.33) are not identically distributed, as for example when contributions come from different atomic species, the single power $C^N(\boldsymbol{\omega})$ is to be replaced by $C^{N_1}(\boldsymbol{\omega})\cdots C^{N_k}(\boldsymbol{\omega})$, where N_j is the number of atoms of the jth atomic species and k is the number of distinct species in the unit cell. Secondly, when the individual steps do not have a zero mean so that $\langle\mathbf{S}\rangle = \boldsymbol{\mu} \neq \mathbf{0}$ then we can still invoke the central limit theorem by replacing the factor $\mathbf{S}$ on the right-hand side of eqn(3.1.44) by $\mathbf{S} - \boldsymbol{\mu}$.

The following sections of this chapter summarize some important approximate pdfs used in crystallographic statistics. Only the lowest-order terms will be given, without correction for non-ideal behaviour.

3.2 The ideal centric and acentric pdfs

A function that plays a most significant role in crystallographic applications is the pdf of the magnitude of the structure factor. This is a powerful indicator of crystallographic symmetry, since it can provide positive evidence that the crystal is centrosymmetric or non-centrosymmetric. Physical properties of crystals, other than their diffraction patterns, may sometimes provide conclusive evidence that the crystal is non-centrosymmetric, but no such evidence on the *presence* of a centre of symmetry can be obtained by these methods.

Let us consider the simplest centrosymmetric space group, $P\bar{1}$. The normalized structure factor is given by

$$E(\mathbf{h}) = 2\sum_{j=1}^{N/2} n_j \cos\theta_j. \qquad (3.2.45)$$

On the assumptions that (i) the atomic contributions to the structure factor are independent, (ii) the atomic coordinates of any atom are rationally independent, and (iii) all the atoms have approximately equal scattering powers, the sum in (3.2.45) can be regarded as a sum of independent and identically distributed random variables. Additionally, in the present discussion we shall also assume that (iv) effects of dispersive scatterers are negligible and (v) there is no non-crystallographic symmetry to be taken into account. Assumption (ii) above implies that all the atoms are located in general Wyckoff positions.

According to the central limit theorem discussed in section 3.1 such a sum is asymptotically normally distributed about the mean

$$\langle E\rangle = 2\sum_{j=1}^{N/2} n_j \langle\cos\theta_j\rangle = 0, \qquad (3.2.46)$$

with a variance

$$\langle E^2\rangle = 4\sum_{j=1}^{N/2} n_j^2 \langle\cos^2\theta_j\rangle = 2\sum_{j=1}^{N/2} n_j^2 = \sum_{j=1}^{N} n_j^2 = 1. \qquad (3.2.47)$$

Note that the results for the mean and the variance are valid for homogeneous as well as heterogeneous atomic compositions. The required approximation to the pdf of E is therefore equal to

$$p_{cen}(E) \simeq \frac{1}{\sqrt{2\pi}} \exp\left(-\frac{E^2}{2}\right). \qquad (3.2.48)$$

Since $E(\mathbf{h})$ is symmetric in $\mathbf{h}$, as may be checked from eqn(3.2.46), the pdf of $|E|$, which is directly related to observed structure amplitude for a centrosymmetric crystal, is

$$p_{cen}(|E|) = 2p_{cen}(E) \simeq \left(\frac{2}{\pi}\right)^{1/2} \exp\left(-\frac{|E|^2}{2}\right). \qquad (3.2.49)$$

Equation (3.2.49) is related to the pdf of $|F|$ (Wilson 1949) by the transformation

$$p(|E|) = \Sigma^{1/2} p(|F|), \tag{3.2.50}$$

where $\Sigma = \sum_{j=1}^{N} f_j^2$. This transformation, as pointed out earlier in section 1.2, approximately eliminates the dependence on scattering angle. Equation (3.2.49) is called the ideal centric pdf. The same result is obtained in section 2.1 as a limiting case of the exact classical formulation of the centric pdf of $|E|$, by assuming that all the atoms are identical and that their number is sufficiently large.

Another well-known useful probability is the cumulative distribution function (cdf) of $|E|$, i.e. the probability that $|E| \leq |E_m|$. This is equivalent to the fraction of $|E|$ values not exceeding $|E_m|$. For centrosymmetric crystals this probability, because of eqn(3.2.49), is given by

$$N_{cen}(|E_m|) \simeq \int_0^{|E_m|} p(|E|)d|E| = \operatorname{erf}\left(\frac{|E_m|}{\sqrt{2}}\right), \tag{3.2.51}$$

where $\operatorname{erf}(x)$ is the error function (e.g. Abramowitz and Stegun, 1972). The use of the cumulative distribution function of the normalized intensity, $z = |E|^2$, was first introduced in intensity statistics by Howells, Phillips and Rogers (1950). Their so-called $N(z)$ test still forms part of most packages of computer programs for automated structure determination by direct methods.

Other statistics of intrinsic value which are also valuable correlates of crystal symmetry include the moments of $p(|E|)$. The even moments of the ideal centric pdf of the magnitude of $|E|$ are given by

$$\langle |E|^{2n} \rangle_{cen} = \sqrt{\frac{2}{\pi}} \int_0^{\infty} |E|^{2n} \exp(-|E|^2/2) d|E| = \frac{(2n)!}{2^n n!}, \tag{3.2.52}$$

and odd moments of the ideal centric pdf of the magnitude of $|E|$ are calculated as

$$\langle |E|^{2n+1} \rangle_{cen} = \sqrt{\frac{2}{\pi}} \int_0^{\infty} |E|^{2n+1} \exp(-|E|^2/2) d|E| = \left(\frac{2}{\pi}\right)^{1/2} 2^n n!. \tag{3.2.53}$$

Some known definite integrals which can be used for these calculations are given in Appendix A.

The ideal acentric pdf and related statistics are derived in a similar manner. The normalized structure factor is now given by $E = A + iB$, where

$$A = \sum_{j=1}^{N} n_j \cos\theta_j \tag{3.2.54}$$

$$B = \sum_{j=1}^{N} n_j \sin\theta_j. \tag{3.2.55}$$

The components A and B are seen to be sums of identically distributed random variables under the same assumptions as in the case of the centric pdf. According to the central limit theorem discussed in section 3.1, both A and B are asymptotically normally distributed about the means

$$\langle A\rangle = \sum_{j=1}^{N} n_j\langle\cos\theta_j\rangle = 0,\ \langle B\rangle = \sum_{j=1}^{N} n_j\langle\sin\theta_j\rangle = 0, \tag{3.2.56}$$

with variances given by

$$\langle A^2\rangle = \sum_{j=1}^{N} n_j^2\langle\cos^2\theta_j\rangle = \frac{1}{2},\ \langle B^2\rangle = \sum_{j=1}^{N} n_j^2\langle\sin^2\theta_j\rangle = \frac{1}{2}. \tag{3.2.57}$$

Hence, the approximate univariate pdfs of A and B are given by

$$g(A) = \frac{1}{\sqrt{\pi}}\exp(-A^2),\text{ and } g(B) = \frac{1}{\sqrt{\pi}}\exp(-B^2). \tag{3.2.58}$$

Since, as is easily verified from eqns (3.2.54) and (3.2.55), A and B are uncorrelated, their joint pdf is given by

$$h(A,B) = g(A)g(B) \simeq \frac{1}{\pi}\exp[-(A^2+B^2)] = \frac{1}{\pi}\exp(-|E|^2). \tag{3.2.59}$$

By transforming A and B to polar coordinates as $A = |E|\cos\varphi$ and $B = |E|\sin\varphi$, where φ is the phase of the structure factor, one sees that the pdf can be represented in terms of the polar coordinates $(|E|,\varphi)$ as

$$r(|E|,\varphi) \simeq \frac{|E|}{\pi}\exp(-|E|^2). \tag{3.2.60}$$

The pdf of $|E|$, $p(|E|)$, is obtained by integrating out the phase φ. By doing this we obtain the ideal acentric pdf of the magnitude of E

$$p_{acen}(|E|) \simeq 2|E|\exp(-|E|^2). \tag{3.2.61}$$

This equation, which is a consequence of the central limit theorem, has already been found in Chapter 2 as a limiting form of the exact acentric pdf of $|E|$ under the assumption that all the atoms are identical and that there are many of them in the unit cell.

The ideal acentric cumulative distribution of $|E|$ is now obtained as

$$N_{acen}(|E_m|) \simeq 2\int_0^{|E_m|} |E|\exp(-|E|^2)d|E| = 1 - \exp(-|E_m|^2). \quad (3.2.62)$$

Finally, the even moments of the ideal acentric pdf of the magnitude of $|E|$ are given by

$$\langle |E|^{2n} \rangle_{acen} \simeq 2\int_0^{\infty} |E|^{2n+1}\exp(-|E|^2)d|E| = n! \quad (3.2.63)$$

and the odd ones are calculated as

$$\langle |E|^{2n+1} \rangle_{acen} \simeq 2\int_0^{\infty} |E|^{2n+2}\exp(-|E|^2)d|E| = \Gamma\left(n + \frac{3}{2}\right) \quad (3.2.64)$$

(see Appendix A). Both even and odd moments of the ideal pdfs of $|E|$ are useful in the resolution of space-group ambiguities, when the central limit theorem can be presumed to be valid. It is mainly the even moments that are required for calculating higher-order approximations, as will be discussed in Chapter 4.

Although the only space groups explicitly considered in this section are $P1$ and $P\bar{1}$ the results obtained here apply to all centrosymmetric and non-centrosymmetric space groups, provided the number of atoms in the unit cell is large and their scattering factors are not too dissimilar. We shall return to this observation and its ramifications in Chapter 5. Statistics of structure-factor magnitude pertaining to crystals or their projections that belong to centrosymmetric or non-centrosymmetric space groups are called *centric* or *acentric* respectively. A summary of some ideal centric and acentric statistics, derived in this section, is presented in Table 3.1.

Table 3.1 Some ideal centric and acentric statistics

	Centric	*Acentric*
pdf of $\|E\|$	$\sqrt{\frac{2}{\pi}}\exp\left(-\frac{\|E\|^2}{2}\right)$	$2\|E\|\exp(-\|E\|^2)$
cdf of $\|E_m\|$	$\mathrm{erf}\left(\frac{\|E_m\|}{\sqrt{2}}\right)$	$1-\exp(-\|E_m\|^2)$
$(2n)$th moment of $\|E\|$	$(2n)!/[2^n n!]$	$n!$
$(2n+1)$th moment of $\|E\|$	$2^{n+\frac{1}{2}}n!/\sqrt{\pi}$	$\Gamma\left(n+\frac{3}{2}\right)$

3.3 Effects of non-crystallographic symmetry

One of the earliest generalizations of Wilson (1949) statistics incorporated the effects of non-crystallographic symmetry into the formalism of intensity statistics based on the central limit theorem. The resulting approximate pdfs of the structure factor will still be termed 'ideal'. Non-crystallographic centrosymmetry was first treated, in the present context, by Lipson and Woolfson (1952) and the results of the latter study were extensively generalized by Rogers and Wilson (1953). We shall confine ourselves, in this section, to deriving some statistical properties of the so-called bicentric distribution in space group $P\bar{1}$. This arises from the presence of a non-crystallographic centre of symmetry in the asymmetric unit of the space group. The representation of the structure factor, appropriate to the above situation, is given by eqn(1.2.16), and is rewritten below in terms of the normalized structure factor

$$E(\mathbf{h}) = 4\cos\delta \sum_{j=1}^{N/4} n_j \cos(\theta_j - \delta), \tag{3.3.65}$$

where $\delta = 2\pi\mathbf{h}^T\mathbf{d}$, $\mathbf{d}$ being the position vector of the non-crystallographic centre of symmetry, and $\theta_j = 2\pi\mathbf{h}^T\mathbf{r}_j$, $\mathbf{r}_j$ being the position vector of the jth atom in the asymmetric subunit (see Fig. 1.1). We assume, as in the previous section, that all the atomic contributions to the structure factor are independent and that δ as well as each of the θ_js are uniformly distributed in $(-\pi, \pi)$. We also assume that the number of the atoms in the unit cell is large and that the scattering factors of the atomic species are not too dissimilar. Effects of anomalous dispersion will also be neglected.

Let us first take δ to be constant, and evaluate the mean and variance of E, with respect to the θ_js alone. The mean, $\langle E\rangle$, is readily seen to vanish and the variance, for a fixed value of δ, is given by

$$\begin{aligned} \sigma_E^2 &= \cos^2\delta \sum_{j=1}^{N/4} 16 n_j^2 \langle\cos^2\theta_j\rangle \\ &= 8\cos^2\delta \sum_{j=1}^{N/4} n_j^2 \\ &= 2\cos^2\delta, \end{aligned} \tag{3.3.66}$$

since $\sum_{j=1}^{N} n_j^2 = 1$. The central limit theorem therefore predicts that the pdf of E, for a given fixed δ, will tend to

$$p_\delta(E) = \frac{1}{(2\pi\sigma_E^2)^{1/2}} \exp\left(-\frac{E^2}{2\sigma_E^2}\right)$$

$$= \frac{1}{(4\pi\cos^2\delta)^{1/2}} \exp\left(-\frac{E^2}{4\cos^2\delta}\right). \tag{3.3.67}$$

The pdf of E is now obtained as an average over all possible δ in eqn(3.3.67). In this way we find

$$p(E) \simeq \frac{2}{\pi}\int_0^{\pi/2} p_\delta(E)d\delta$$
$$= \frac{1}{\pi^{3/2}}\int_0^{\pi/2} \sec\delta \exp\left(-\frac{E^2}{4}\sec^2\delta\right) d\delta. \tag{3.3.68}$$

If we change the variable to $\rho = \sec^2\delta$, the integral in (3.3.68) is transformed to

$$p(E) \simeq \frac{1}{2\pi^{3/2}}\int_1^{\infty} \frac{\exp(-E^2\rho/4)}{\sqrt{\rho^2-\rho}} d\rho. \tag{3.3.69}$$

A further change of the variable, $\alpha = \rho - 1$, leads to

$$p(E) \simeq \frac{\exp(-E^2/4)}{2\pi^{3/2}}\int_0^{\infty} \frac{\exp(-E^2\alpha/4)}{\sqrt{\alpha(\alpha+1)}} d\alpha. \tag{3.3.70}$$

The integral in eqn(3.3.70) can be evaluated exactly [see Appendix A, eqn(A.0.4)], which leads to

$$p(E) \simeq \frac{1}{2\pi^{3/2}} \exp\left(-\frac{E^2}{8}\right) K_0\left(\frac{E^2}{8}\right), \tag{3.3.71}$$

where $K_0(x)$ is a modified Bessel function of the second kind and of zeroth order. The pdf of the magnitude of E, $p(|E|)$, is just $p_{bicen}(|E|) = 2p(E)$. This is the so-called bicentric pdf of $|E|$, derived by Rogers and Wilson (1953) and represented here in terms of the magnitude of the normalized structure factor. Generalizations of this pdf, and a derivation of the exact form of the pdf, will be discussed in Chapters 4 and 5. It should be pointed out, however, that eqn(3.3.71) is accurate when applied to structures that contain a reasonably large number of not too dissimilar atoms in the asymmetric subunit of the bicentric arrangement.

Another important ideal bicentric statistical property is summarized in the cumulative distribution function,

$$N_{bicen}(|E_m|) = \int_0^{|E_m|} p_{bicen}(|E|)d|E|,$$

which can be obtained by numerical integration of (3.3.71). Figure 3.1 shows the cumulative distribution functions of $|E|$ for the acentric, centric and bicentric distributions.

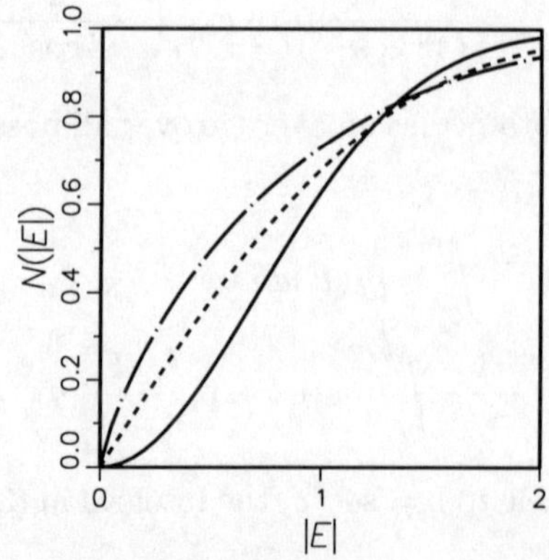

FIG. 3.1. *Acentric (solid), centric (dashed) and bicentric (dashed-dotted) cumulative distribution functions of* $|E|$.

The moments of $p_{bicen}(|E|)$ can be obtained in closed form in the approximation implied by the central limit theorem. For the even moments we have

$$\langle |E|^{2n} \rangle \simeq \frac{1}{\pi^{3/2}} \int_0^\infty |E|^{2n} \exp\left(-\frac{|E|^2}{8}\right) K_0\left(\frac{|E|^2}{8}\right) d|E|. \qquad (3.3.72)$$

If we change the variable to $x = |E|^2$, the integral in (3.3.72) transforms to

$$\langle |E|^{2n} \rangle \simeq \frac{1}{2\pi^{3/2}} \int_0^\infty x^{n-\frac{1}{2}} \exp\left(-\frac{x}{8}\right) K_0\left(\frac{x}{8}\right) dx = \frac{[(2n)!]^2}{2^{2n}(n!)^3}. \qquad (3.3.73)$$

[A relevant known definite integral is given in Appendix A, eqn(A.0.5).] We shall return to these so-called ideal statistics in comparing them with generalized and exact ones in later chapters.

3.4 The subcentric distribution in the Gaussian approximation

In this section we consider the calculation of an ideal pdf, for two completely unrelated systems: (i) a $P\overline{1}$ structure with dispersive scatterers in the asymmetric unit (Wilson 1980), and (ii) a non-centrosymmetric (space group $P1$) structure with centrosymmetric fragments in the asymmetric unit of the space group (Srinivasan and Parthasarathy 1976). These two configurations have in common the fact that there are unequal bounds on the real and imaginary parts of the structure factor. While this fact is not significant for the central limit theorem analysis, it will be taken into account in our derivation of exact representations of these pdfs (section 5.8).

In this section we derive these ideal pdfs and interpret their parameters for the two systems just mentioned.

If dispersion is to be accounted for, the conventional atomic scattering factor is represented as a complex quantity $f_j = f'_j + i f''_j$. The corresponding average intensity then becomes

$$\Sigma = \sum_{j=1}^{N} |f_j|^2 = \sum_{j=1}[(f'_j)^2 + (f''_j)^2] \tag{3.4.74}$$

and the normalized scattering factor is given by $n_j = n'_j + i n''_j$, where

$$n'_j = \frac{f'_j}{\Sigma^{1/2}} \text{ and } n''_j = \frac{f''_j}{\Sigma^{1/2}}. \tag{3.4.75}$$

It follows that $\sum_{j=1}^{N} |n_j|^2 = 1$.

The normalized structure factor for the space group $P\bar{1}$ is given by

$$E = 2\sum_{j=1}^{N/2}(n'_j + i n''_j)\cos\theta_j \equiv A + iB. \tag{3.4.76}$$

Under the assumption that the θs are uniformly distributed the means of A and B vanish, and the variances of these quantities are

$$\sigma_A^2 = 4\sum_{j=1}^{N/2}(n'_j)^2\langle\cos^2\theta_j\rangle = \sum_{j=1}^{N}(n'_j)^2 \tag{3.4.77}$$

and, analogously,

$$\sigma_B^2 = \sum_{j=1}^{N}(n''_j)^2. \tag{3.4.78}$$

However, the derivation of an approximate pdf cannot proceed exactly as in the case of the ideal acentric pdf because A and B are correlated. The correlation function is

$$\begin{aligned}\langle AB\rangle &= 4\sum_{j=1}^{N/2} n'_j n''_j \langle\cos^2\theta\rangle \\ &= \sum_{j=1}^{N} n'_j n''_j.\end{aligned} \tag{3.4.79}$$

Our derivation will follow in broad outline that of Wilson (1980). We shall first make use of section 3.1 to find the joint pdf of A and B and, finally, the required ideal pdf, $p(|E|)$, which will then account for effects of dispersion.

Recall that the last part of section 3.1 deals with an application of the central limit theorem to multivariate problems with correlated random variables. If in eqn(3.1.42) we replace S_1 by A, S_2 by B, $N\langle X^2\rangle$ by $N(n')^2 \equiv \sigma_A^2$ and $N\langle (X')^2\rangle$ by $N(n'')^2 \equiv \sigma_B^2$ we obtain the joint pdf of A and B for the equal-atom dispersive case as

$$p(A,B) \simeq \frac{1}{2\pi\sigma_A\sigma_B(1-\rho^2)^{1/2}} \exp\left[-\frac{2(\sigma_B^2 A^2 - 2\rho\sigma_A\sigma_B AB + \sigma_A^2 B^2)}{4\sigma_A^2\sigma_B^2(1-\rho^2)}\right], \tag{3.4.80}$$

where ρ is defined in eqn(3.1.41). Of course, if the atoms are not of exactly the same type, i.e. in the case of small or moderate atomic heterogeneity, the more general expressions, eqn(3.4.77) and eqn(3.4.78), for σ_A^2 and σ_B^2 can be used with eqn(3.4.80) and the pdfs that are based on it.

In order to derive the required pdf of $|E|$ we transform to polar coordinates by setting $A = |E|\cos\varphi$ and $B = |E|\sin\varphi$. In the present treatment the phase angle φ is obviously *not* uniformly distributed over $(-\pi, \pi)$. However, before carrying out the algebra and integration it is convenient to consider the summation

$$\mathbf{S_E} = \sum_{j=1}^{N}(n'_j + in''_j)^2 = \sigma_A^2 - \sigma_B^2 + 2i\rho\sigma_A\sigma_B. \tag{3.4.81}$$

The magnitude of $\mathbf{S_E}$ is

$$S_E = [(\sigma_A^2 - \sigma_B^2)^2 + 4\rho^2\sigma_A^2\sigma_B^2]^{1/2} \tag{3.4.82}$$

$$= [(\sigma_A^2 + \sigma_B^2)^2 - 4\sigma_A^2\sigma_B^2(1-\rho^2)]^{1/2}. \tag{3.4.83}$$

Since eqns (3.4.75), (3.4.77) and (3.4.78) lead to the identity $\sigma_A^2 + \sigma_B^2 = 1$, we can write this more succinctly as

$$S_E = (1-\mu^2)^{1/2}, \tag{3.4.84}$$

where

$$\mu = 2\sigma_A\sigma_B(1-\rho^2)^{1/2}, \tag{3.4.85}$$

and the phase of $\mathbf{S_E}$ is obtained from eqn(3.4.81) as

$$\delta = \tan^{-1}\frac{2\rho\sigma_A\sigma_B}{\sigma_A^2 - \sigma_B^2}. \tag{3.4.86}$$

It follows that $2\rho\sigma_A\sigma_B = S_E\sin\delta$ and $\sigma_A^2 - \sigma_B^2 = S_E\cos\delta$, where δ is a constant angle. If we now substitute in (3.4.80) the expressions for A and B in terms of the magnitude $|E|$ and phase φ of the normalized structure factor we obtain

$$p(A,B) = \frac{1}{\pi\mu} \exp\left[-\frac{2|E|^2(\sigma_B^2 \cos^2\varphi - 2\rho\sigma_A\sigma_B \cos\varphi \sin\varphi + \sigma_A^2 \sin^2\varphi)}{\mu^2}\right]. \tag{3.4.87}$$

We further observe that the change of the variable from (A, B) to $(|E|, \varphi)$ requires a multiplication of (3.4.87) by the Jacobian which in this case is just $|E|$. If we make use of the identities $\cos^2(a) = (1 + \cos 2a)/2$ and $\sin^2(a) = (1 - \cos 2a)/2$ we can bring the desired pdf to the form

$$p(|E|) \simeq \frac{|E|}{\pi\mu} \exp\left(-\frac{|E|^2}{\mu^2}\right) \int_{-\pi}^{\pi} \exp\left[\frac{S_E|E|^2}{\mu^2} \cos(2\varphi - \delta)\right] d\varphi. \tag{3.4.88}$$

The integral can be evaluated as in (3.1.29), leading to the final result

$$p(|E|) \simeq \frac{2|E|}{\mu} \exp\left(-\frac{|E|^2}{\mu^2}\right) I_0\left(\frac{S_E|E|^2}{\mu^2}\right), \tag{3.4.89}$$

in agreement with eqn(37) of Wilson (1985). An analogous result is presented by Srinivasan and Parthasarathy (1976) for the case of partial non-crystallographic centrosymmetry in the asymmetric unit of space group $P1$. In the case of dispersion, S_E is given by

$$S_E = \frac{\left|\sum_{j=1}^{N}(f_j' + if_j'')^2\right|}{\sum_{j=1}^{N}|f_j' + if_j''|^2},$$

while in the case of partial centrosymmetry the parameter equivalent to S_E is the ratio Σ_c/Σ, where Σ_c is the sum of the squares of the scattering factors of the atoms occupying the centrosymmetric fragment (Srinivasan and Parthasarathy 1976; Shmueli and Wilson 1983).

Let us now consider the two important limiting cases. Since $\mu = (1 - S_E)^{1/2}$, when S_E tends to zero eqn(3.4.89) reduces immediately to the ideal acentric pdf given by eqn(3.2.61). This is so in the case of dispersion when the real and imaginary parts of the structure factor are uncorrelated, and the phase of the atomic scattering factor is uniformly distributed over $(-\pi, \pi)$. It is an obvious limiting case for partial centrosymmetry when the centrosymmetric fragment is empty. The second limiting case is one in which S_E tends to unity. One expects that in this case the pdf should tend to the ideal centric pdf, given by eqn(3.2.49). This corresponds to the absence of dispersive scatterers (in $P\bar{1}$) and to the whole asymmetric unit of the space group $P1$ being centrosymmetric. However, the mathematical limiting process, which is used to demonstrate the final result, is less straightforward. The following derivation of the moments of eqn(3.4.89), close to that briefly presented by Shmueli and Wilson (1983) and to which we shall refer in the next chapter, will be helpful in carrying out this limiting process.

The even moment of the pdf given by eqn(3.4.89) is given by

$$\langle |E|^{2n} \rangle \simeq 2\alpha^{1/2} \int_0^\infty |E|^{2n+1} \exp(-\alpha |E|^2) I_0(\beta |E|^2) d|E|, \qquad (3.4.90)$$

where $\alpha = 1/\mu^2$ and $\beta = S_E/\mu^2$. If we change the variable from the magnitude of the normalized structure factor to the normalized intensity, $z = |E|^2$, the integral (3.4.90) is transformed to

$$\langle z^n \rangle \simeq \alpha^{1/2} \int_0^\infty z^n \exp(-\alpha z) I_0(\beta z) dz \qquad (3.4.91)$$

$$= \frac{n! \alpha^{1/2}}{(\alpha^2 - \beta^2)^{(n+1)/2}} P_n \left(\frac{\alpha}{\sqrt{\alpha^2 - \beta^2}} \right), \qquad (3.4.92)$$

where $P_n(x)$ is a Legendre polynomial (Magnus, Oberhettinger and Soni 1966). If we note that

$$\alpha^2 - \beta^2 = \alpha^2(1 - S_E) = \alpha^2 \mu^2 = \alpha, \qquad (3.4.93)$$

eqn(3.4.92) reduces to

$$\langle z^n \rangle \simeq n! \alpha^{-n/2} P_n(\alpha^{1/2}), \qquad (3.4.94)$$

which is the result given by Shmueli and Wilson (1983). Equation (3.4.94) will be useful in constructing orthogonal polynomials for the higher approximations to be discussed in the next chapter. In the present context, let us write the Legendre polynomial explicitly, and construct a series expression for the nth moment of z. If we make use of eqn(3.4.94) and an explicit expression for the Legendre polynomial $P_n(x)$ (e.g. from Magnus, Oberhettinger and Soni 1966; section 5.4.2) we obtain

$$\langle z^n \rangle \equiv \langle |E|^{2n} \rangle \simeq \frac{(2n)!}{2^n n!} \times \left[1 - \frac{n(n-1)}{2(2n-1)} \alpha^{-1} + O(\alpha^{-2}) - \cdots \right]. \qquad (3.4.95)$$

Since $\alpha = 1/(1 - S_E^2)$, one sees that as S_E tends to unity, the parameter α tends to infinity, and all the terms in the brackets on the right-hand side of (3.4.95), except the first one, tend to zero. The limiting value of the even moment of $|E|$ is therefore $\frac{(2n)!}{2^n n!}$, as found for the even moments of the ideal centric pdf in section 3.2. The ideal subcentric pdf and its moments will be applied to the construction of higher approximations to the Gaussian density in Chapter 4. Problems relating to dispersion and partial centrosymmetry will be reconsidered in Chapter 5, which deals with derivations of exact pdfs for $|E|$.

3.5 Conditional density of a three-phase invariant

Early studies of the solution of the phase problem focussed on quantities which are structure invariant, i.e. do not depend on the choice of the space-group origin. These were first formally defined by Hauptman and Karle in their monograph (1953). A structure invariant of great importance in direct methods is the well-known three-phase invariant, which can take the forms $\varphi_{\mathbf{h}} + \varphi_{\mathbf{k}} + \varphi_{-\mathbf{h}-\mathbf{k}}$, $\varphi_{-\mathbf{h}} + \varphi_{\mathbf{k}} + \varphi_{\mathbf{h}-\mathbf{k}}$ or any combination of three phases for which the sum of the diffraction vectors is equal to zero. A conditional pdf of the phase $\varphi_{\mathbf{h}}$, given the phases $\varphi_{\mathbf{k}}$ and $\varphi_{\mathbf{h}-\mathbf{k}}$, and the magnitudes $|E_{\mathbf{h}}|$, $|E_{\mathbf{k}}|$ and $|E_{\mathbf{h}-\mathbf{k}}|$ was first derived by Cochran (1955) on the basis of the central limit theorem. It is a simple and at the same time most important example of an application of this theorem to multivariate structure-factor statistics, which are indispensable in the use of direct methods of phase determination. We shall first rederive this conditional pdf for space group $P1$ and then show its relationship to the three-phase structure invariant.

We shall assume in what follows that all the atoms are identical and write the normalized structure factor as

$$E_{\mathbf{h}} = \frac{1}{\sqrt{N}} \sum_{j=1}^{N} \exp(2\pi i \mathbf{h}^T \mathbf{r}_j) \tag{3.5.96}$$

$$= \frac{1}{\sqrt{N}} \sum_{j=1}^{N} \exp(2\pi i \mathbf{k}^T \mathbf{r}_j) \exp[2\pi i (\mathbf{h} - \mathbf{k})^T \mathbf{r}_j] \tag{3.5.97}$$

$$\equiv \frac{1}{\sqrt{N}} \sum_{j=1}^{N} (\xi_j + i\eta_j) \tag{3.5.98}$$

$$\equiv A_{\mathbf{h}} + iB_{\mathbf{h}},$$

where N, the number of atoms in the unit cell, is reasonably large. We further assume, following Woolfson (1954), that the vector $\mathbf{h}$ is fixed while $\mathbf{k}$ in eqn(3.5.97) ranges uniformly over those diffraction vectors, for which $E_{\mathbf{k}}$ and $E_{\mathbf{h}-\mathbf{k}}$ have specified magnitudes and phases (note that $E_{-\mathbf{h}} E_{\mathbf{k}} E_{\mathbf{h}-\mathbf{k}}$ is a structure invariant). This assumption will lead to the desired conditioning of the joint pdf of $A_{\mathbf{h}}$ and $B_{\mathbf{h}}$ on the other members of the triplet. If, in addition, the atoms are randomly distributed throughout the unit cell, the averages of the expression $\exp(2\pi i \mathbf{k}^T \mathbf{r}_j)$ taken over atomic position vectors $\mathbf{r}_j$ or over diffraction vectors $\mathbf{k}$ are approximately equivalent, at least for large N, since in either of them the scalar product $\mathbf{k}^T \mathbf{r}_j$ can be taken as uniformly distributed over (0,1).

According to the central limit theorem $A_{\mathbf{h}}$ tends to be normally distributed with mean

$$\langle A_{\mathbf{h}} \rangle = N^{-1/2} \sum_{j=1}^{N} \langle \xi_j \rangle \tag{3.5.99}$$

and variance

$$\sigma^2(A_{\mathbf{h}}) = N^{-1}\sum_{j-1}^{N}\left[\langle\xi_j^2\rangle - \langle\xi_j\rangle^2\right]. \tag{3.5.100}$$

We again assume that our random variables are independent. However, in contrast to the examples of earlier sections the mean $\langle A_{\mathbf{h}}\rangle$ does not vanish. If we use the representation of $E_{\mathbf{h}}$ given by eqn(3.5.97) along with the assumptions specified in this section, and the relation

$$\langle\exp(2\pi i\mathbf{k}^T\mathbf{r}_j\rangle_{\mathbf{r}} = \frac{1}{N}\sum_{j=1}^{N}\exp(2\pi i\mathbf{k}^T\mathbf{r}_j) = N^{-1/2}E_{\mathbf{k}}$$

we obtain, after some calculation,

$$\begin{aligned}\langle A_{\mathbf{h}}\rangle &= N^{-1/2}\Re(E_{\mathbf{k}}E_{\mathbf{h-k}})\\ &= N^{-1/2}|E_{\mathbf{k}}E_{\mathbf{h-k}}|\cos(\varphi_{\mathbf{k}}+\varphi_{\mathbf{h-k}})\end{aligned} \tag{3.5.101}$$

and

$$\sigma^2(A_{\mathbf{h}}) = \frac{1}{2} - \frac{|E_{\mathbf{k}}E_{\mathbf{h-k}}|^2\cos^2(\varphi_{\mathbf{k}}+\varphi_{\mathbf{h-k}})}{N^2} \tag{3.5.102}$$

$$\simeq \frac{1}{2} \tag{3.5.103}$$

for sufficiently large N. Similarly, we obtain

$$\langle B_{\mathbf{h}}\rangle = N^{-1/2}|E_{\mathbf{k}}E_{\mathbf{h-k}}|\sin(\varphi_{\mathbf{k}}+\varphi_{\mathbf{h-k}}) \tag{3.5.104}$$

with the value of 1/2 as an approximation for the variance of $B_{\mathbf{h}}$. If $A_{\mathbf{h}}$ and $B_{\mathbf{h}}$ are also assumed to be uncorrelated, their joint conditional pdf is given by

$$p(A_{\mathbf{h}}, B_{\mathbf{h}}|E_{\mathbf{k}}, E_{\mathbf{h-k}}) \simeq \frac{1}{\pi}\exp\{-[(A_{\mathbf{h}}-\langle A_{\mathbf{h}}\rangle)^2+(B_{\mathbf{h}}-\langle B_{\mathbf{h}}\rangle)^2]\}, \tag{3.5.105}$$

where the Es occur in $\langle A_{\mathbf{h}}\rangle$ and $\langle B_{\mathbf{hh}}\rangle$ as specified in eqns (3.5.101) and (3.5.104). As pointed out by Cochran (1955), when $\langle E_{\mathbf{h}}\rangle = 0$ eqn(3.5.105) reduces to the form of the Wilson (1949) acentric pdf, *before the phase of the structure factor has been integrated out* (see section 3.2). This observation is highly relevant to the evolution of direct methods from intensity statistics.

Since the change of the variables $(A_{\mathbf{h}}, B_{\mathbf{h}})$ to $(|E_{\mathbf{h}}|, \varphi_{\mathbf{h}})$ involves only a multiplication of eqn(3.5.105) by the Jacobian $|E|$, we have from eqns (3.5.101), (3.5.104) and (3.5.105)

$$p(|E_{\mathbf{h}}|, \varphi_{\mathbf{h}}|E_{\mathbf{k}}, E_{\mathbf{h-k}}) \simeq \frac{|E_{\mathbf{h}}|}{\pi}\exp[-W+\kappa\cos(\varphi_{\mathbf{h}}-\varphi_{\mathbf{k}}-\varphi_{\mathbf{h-k}})], \tag{3.5.106}$$

where

$$W = |E_{\mathbf{h}}|^2 + \frac{|E_{\mathbf{k}}E_{\mathbf{h-k}}|}{\sqrt{N}}$$

and

$$\kappa = \frac{2}{\sqrt{N}}|E_{\mathbf{h}}E_{\mathbf{k}}E_{\mathbf{h-k}}|. \tag{3.5.107}$$

The normalized conditional pdf of $\varphi_{\mathbf{h}}$ alone is given by

$$p(\varphi_{\mathbf{h}}|\cdots) \simeq \frac{\exp[\kappa\cos(\varphi_{\mathbf{h}} - \varphi_{\mathbf{k}} - \varphi_{\mathbf{h-k}})]}{\int_{-\pi}^{\pi}\exp[\kappa\cos(\varphi_{\mathbf{h}} - \varphi_{\mathbf{k}} - \varphi_{\mathbf{h-k}})]d\varphi_{\mathbf{h}}} \tag{3.5.108}$$

and if we make use of the integral representation of the modified Bessel function of the first kind, eqn(3.5.108) reduces to

$$p(\varphi_{\mathbf{h}}|\cdots) \simeq \frac{1}{2\pi I_0(\kappa)}\exp[\kappa\cos(\varphi_{\mathbf{h}} - \varphi_{\mathbf{k}} - \varphi_{\mathbf{h-k}})]. \tag{3.5.109}$$

This is known as Cochran's equation. Its applications to various stages of phase determination are described in detail in the crystallographic literature (e.g. Giacovazzo 1980).

Since, in the non-centrosymmetric and dispersionless case $\varphi_{-\mathbf{h}} = -\varphi_{\mathbf{h}}$ for any acentric reflection, it is clear that the argument of the cosine in eqn(3.5.109) is a structure invariant. If we denote the three-phase invariant $\varphi_{\mathbf{h}} - \varphi_{\mathbf{k}} - \varphi_{\mathbf{h-k}} = \varphi_{\mathbf{h}} + \varphi_{-\mathbf{k}} + \varphi_{\mathbf{k-h}}$ by Φ, Cochran's equation can be rewritten as the conditional probability density of the three-phase invariant which has the form

$$p(\Phi||E_{\mathbf{h}}|, |E_{\kappa}|, |E_{-\mathbf{h-k}}|) \simeq \frac{1}{2\pi I_0(\kappa)}\exp(\kappa\cos\Phi). \tag{3.5.110}$$

This important conditional density is to be considered as a zero-order approximation. We shall generalize it by deriving an exact formula by the Fourier method in Chapter 6, showing there that our exact expression reduces to the central limit theorem result in the appropriate limit.

4

HIGHER-ORDER APPROXIMATIONS

4.1 Mathematical background

4.1.1 *Expansion in orthogonal polynomials*

We have shown that the central limit theorem can be used to calculate useful approximations to both univariate and multivariate pdfs that are generally easy to evaluate numerically. However, there are circumstances in which Gaussian approximations may be qualitatively inaccurate. Under these circumstances it is obviously desirable to find corrections to the Gaussian approximation or, alternatively, to find representations of the pdfs that are exact and whose numerical implementations are not too difficult. Both approaches to this goal will be discussed, Chapter 4 being largely devoted to approximate correction methods that build on the Gaussian approximation. However, before discussing the nature of these corrections we present an example in which the central limit theorem is shown to lead to a poor approximation to the pdf for the intensity.

Let us suppose that we want to calculate an approximation to the pdf of $E(\mathbf{h})$ for a crystal whose structure belongs to space group $P\bar{1}$. In this crystal the asymmetric unit consists of $N/2$ atoms, $N/2 - 1$ having the same atomic number, while the remaining one has a considerably greater atomic number. The representation of $E(\mathbf{h})$ for this space group is given in eqn(1.4.21) which we repeat here for convenience:

$$E(\mathbf{h}) = 2\sum_{j=1}^{N/2} n_j \cos\theta_j. \tag{4.1.1}$$

Our earlier assumption on the composition of the unit cell is equivalent to setting $n_1 = n_2 = \cdots = n_{N/2-1} = n$ and $n_{N/2} = n'$ where $n' \gg n$, which allows us to write

$$E(\mathbf{h}) = 2n \sum_{j=1}^{N/2-1} \cos\theta_j + 2n' \cos\theta_{N/2}. \tag{4.1.2}$$

When there is considerable atomic heterogeneity and N is not too large, it is natural to expect that the pdf of $E(\mathbf{h})$ will be dominated by the contribution of the heavy atom.

To determine the qualitative behaviour of the pdf of E, $p(E)$, we consider the pdf of E for a structure containing just the single heavy atom in the asymmetric unit, and neglect the contribution of the light atoms. The normalized structure factor is now approximately given by

$$E(\mathbf{h}) \simeq 2n' \cos\theta. \tag{4.1.3}$$

The exact pdf corresponding to this expression for $E(\mathbf{h})$ is found by first calculating the characteristic function

$$C(\omega) = \langle \exp[i\omega_1(2n' \cos\theta)] \rangle = J_0(2n'\omega_1). \tag{4.1.4}$$

The pdf of E is found by inverting the Fourier transform

$$p(E) = \frac{1}{2\pi} \int_{-\infty}^{\infty} J_0(2n'\omega_1) \exp(-i\omega_1 E) d\omega_1,$$

leading to the value

$$p(E) = \begin{cases} \frac{1}{\pi} \frac{1}{\sqrt{4(n')^2 - E^2}} & E^2 < 4(n')^2 \\ 0 & E^2 > 4(n')^2. \end{cases} \tag{4.1.5}$$

Thus, $p(E)$ is infinite at $E = 2n'$ and has a minimum at $E = 0$. This behaviour differs qualitatively from that described by a Gaussian pdf, whose maximum occurs at $E = 0$. When the total number of atoms in the unit cell is not too large, we expect that the effect of the remaining $N/2 - 1$ light atoms will be to perturb the form of $p(E)$ by reducing the two infinities of $p(E)$ to a finite value but not wiping out the maxima at the endpoints. The pdf shown in Figure 4.1, which consists of a histogram generated from simulated data, indicates that this qualitative argument is in fact an accurate one. The distribution shown in the figure was obtained by constructing a large set of values of $|E|$, as defined by eqn(4.1.1), where the atomic phase angle is being replaced, upon each occurrence, by 2π times a random number uniformly distributed over the interval (0,1). The set of the $|E|$ values simulated in this way is then sorted into a frequency histogram which is the one shown in the figure. This is a close approximation to a pdf, allowing us to compare theoretical pdfs to these 'experimental' results. The asymmetric unit in our example is assumed to contain 14 carbon atoms and one uranium atom. The discrepancy between the simulated histogram and the Gaussian in Fig. 4.1 illustrates the departure of the pdf from the approximation furnished by the central limit theorem. In this chapter we indicate a number of techniques which can be used to produce corrections to the lowest-order Gaussian approximation.

In order to find these corrections to the pdf of the structure factor we need first to know what further parameters will be needed in addition to

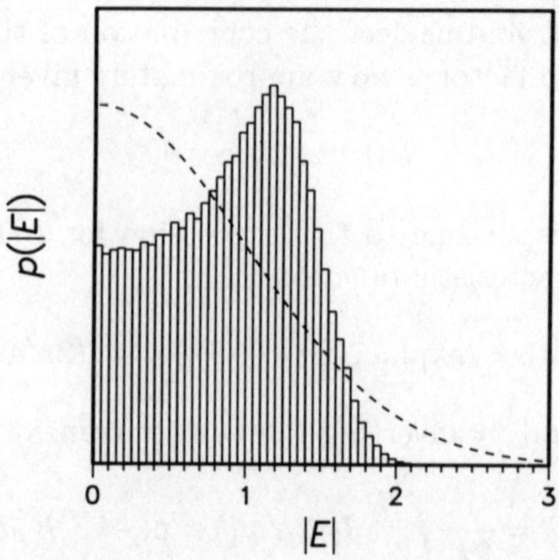

FIG. 4.1. *Departure from the ideal pdf caused by the presence of an outstandingly heavy atom.* The dashed line represents a Gaussian pdf calculated using the central limit theorem.

the first two moments required for the derivation of the Gaussian. The additional information almost universally used consists of moments of orders higher than the second. For low symmetries these can be calculated in a rather straightforward fashion. Some early sources in the crystallographic literature developing this theme are contained in articles by Karle and Hauptman (1953), Hauptman and Karle (1953*a*), Bertaut (1955) and Klug (1958). The best known mathematical technique to generate these corrections is based on the use of orthogonal polynomials (Szegö 1939). Since this technique is not necessarily confined to deriving corrections to a Gaussian pdf, we develop the idea in a more general setting. This will allow us to derive corrections to the pdf of the magnitude of the structure factor even when the lowest-order approximation is not a Gaussian (see, for example, Shmueli and Wilson 1983).

For simplicity we restrict this discussion to one dimension. Suppose that we find some lowest-order approximation, $p_0(x)$, to the (unknown) exact pdf, $p(x)$. Let $\{\phi_k(x)\}$, $k = 0, 1, 2, \ldots$, be the set of polynomials orthogonal with respect to $p_0(x)$, in the sense that

$$\int_a^b \phi_m(x)\phi_n(x)p_0(x)dx = \delta_{mn}, \tag{4.1.6}$$

where $\delta_{mn} = 0$ when $m \neq n$ and $\delta_{mn} = 1$ when $m = n$, and (a, b) is the range of x relevant for the particular physical problem. If the set $\{\phi_k(x)\}$ is complete, that is, if the infinite set of these polynomials can be used to expand an arbitrary function (Szegö, 1939) then it is natural to expand $p(x)$ in a series of the form

$$p(x) = p_0(x) \sum_{k=0}^{\infty} d_k \phi_k(x). \tag{4.1.7}$$

In this expansion the d_k are a set of constants each of which, we will show, is expressible as a linear combination of the moments of $p(x)$. The complete set of these constants accounts for the discrepancy between $p_0(x)$ and $p(x)$. A demonstration of this fact is a consequence of the orthogonality property expressed by eqn(4.1.6). Let the mth polynomial be written as $\phi(x) = \sum_j c_{mj} x^j$. An explicit representation of a coefficient d_m is then

$$d_m = \int_a^b \phi_m(x) p(x) dx = \langle \phi_m \rangle = \sum_{j=0}^{m} c_{mj} \langle x^j \rangle, \tag{4.1.8}$$

where $\langle x^j \rangle$ is the jth moment of the *exact* pdf $p(x)$. If the entire sets of d_k and $\langle x^j \rangle$ were known and it were possible to sum the infinite set of terms in (4.1.7) then one would, in principle, have an exact representation of $p(x)$. In practice, the expansion in eqn(4.1.7) is truncated after a small number of terms, because of their rapidly increasing complexity, which therefore furnishes an approximation to the correction term rather than an exact representation. We note that the truncated approximation to the exact pdf is not necessarily non-negative over the entire range. However, the discrepancies usually occur in the tail of the curve where the value of the pdf is small in any case.

To summarize, the two main steps required for calculating an approximation are:

1. Calculate the set of polynomials orthogonal with respect to $p_0(x)$.
2. Calculate the exact moments of $p(x)$.

Notice that one has to calculate the exact moments without prior knowledge of $p(x)$. This is feasible in crystallographic applications because it is possible to evaluate the moments of the structure factor, or its real and imaginary parts, in terms of moments of the corresponding trigonometric structure factors (tsfs). The first extensive account of such calculations was given by Foster and Hargrqeaves (1963) and the subject was further developed by Wilson (1978), who calculated the fourth moment of the tsf for nearly all space groups. Shmueli and Kaldor (1981,1983) devised computer-oriented algorithms and applied them to the calculation of the first few even moments of the tsfs for all space groups and Shmueli (1982) gave closed expressions for even moments of tsf of any order for most space groups of symmetries not higher than orthorhombic. Some details on these recent developments are given in section 4.2.

The construction of orthogonal polynomials to be used with eqn(4.1.7) is a simpler matter. They can be generated recursively in terms of moments of $p_0(x)$, provided the latter exist. To see how this is done, let the kth moment of $p_0(x)$ be denoted by $\langle x^k \rangle_0$, and set $\phi_0(x) = 1$. This clearly satisfies eqn(4.1.6) since

$$\int_a^b p_0(x)dx = 1. \tag{4.1.9}$$

To determine the next orthogonal polynomial make the assumption that $\phi_1(x)$ is a linear function of x

$$\phi_1(x) = U + Vx, \tag{4.1.10}$$

where U and V are constants which will be found from the orthogonality relation in eqn(4.1.6). That equation implies that when $m = 1$ and $n = 0$ we have

$$U + V\langle x \rangle_0 = 0. \tag{4.1.11}$$

This furnishes one relation between U and V. A second relation is required to determine both parameters. This may be found from (4.1.7) by setting $m = n = 1$, which is equivalent to

$$U^2 + 2UV\langle x \rangle_0 + V^2\langle x^2 \rangle_0 = 1. \tag{4.1.12}$$

On substituting $U = -V\langle x \rangle_0$ found from eqn(4.1.11) into this equation we find a single linear equation for V, whose solution is

$$V = \frac{1}{\sigma}, \tag{4.1.13}$$

in which $\sigma^2 \equiv \langle x^2 \rangle_0 - \langle x \rangle_0^2$. The remaining parameter, U, is then found from eqn(4.1.11) to be $U = -\langle x \rangle_0/\sigma$.

This procedure for finding orthogonal polynomials, usually termed the Gram–Schmidt algorithm (e.g. Strang 1988), can be continued by assuming that $\phi^2(x)$ is quadratic in x:

$$\phi_2(x) = U' + V'x + W'x^2. \tag{4.1.14}$$

The constants, U', V' and W', are found by requiring that $\phi_2(x)$ be orthogonal to $\phi_0(x)$ and $\phi_1(x)$ and that its norm satisfies

$$\int_a^b [\phi_n(x)]^2 p_0(x)dx = 1, \quad n = 0, 1, 2, 3\ldots. \tag{4.1.15}$$

While successive polynomials can be generated in this way, the algebra quickly becomes tedious. A general expression can be found for $\phi_n(x)$ in terms of a determinant (Szegö 1939). For simplicity write $\mu_n = \langle x^n \rangle_0$. The result given by Szegö for a polynomial of the order $2n$ can be expressed as

$$\phi_n(x) = K_n \begin{vmatrix} x^{2n} & x^{2n-2} & \cdot & x^2 & 1 \\ \mu_{2n-1} & \mu_{2n-2} & \cdot & \mu_n & \mu_{n-1} \\ \cdot & \cdot & \cdot & \cdot & \cdot \\ \mu_{n+1} & \mu_n & \cdot & \mu_2 & \mu_1 \\ \mu_n & \mu_{n-1} & \cdot & \mu_1 & \mu_0 \end{vmatrix}, \tag{4.1.16}$$

where K_n is a normalization constant that must be found from eqn(4.1.15). We can therefore express the coefficients c_{mj} in eqn(4.1.8) in terms of the (usually) available moments of $p_0(x)$ allowing us to rewrite the expansion in eqn(4.1.7) as

$$p(x) = p_0(x) \sum_{m=0}^{\infty} \left(\sum_{j=0}^{m} c_{mj} \langle x^j \rangle \right) \phi_m(x). \tag{4.1.17}$$

An approximation to the correction, i.e. all of the terms exclusive of $p_0(x)$, is generally found by truncating the series in this last equation at a finite value of m determined by the number of available moments. When the discrepancy between the exact $p(x)$ and the original approximation $p_0(x)$ is not too large, these few terms may often suffice for practical applications as we show later by a number of examples. Symbolic programming (i.e. the use of languages like REDUCE or *Mathematica*) is useful for carrying out error-free calculations of the correction factors since these often involve cumbersome and error-prone, albeit straightforward, algebraic manipulations.

Several examples of such correction factors are used in crystallographic applications. We have seen, in the last chapter, that the lowest-order approximation to the pdf of E for a centrosymmetric crystal is a Gaussian. In the terms used in this section we have

$$p_0(E) \simeq \frac{1}{\sqrt{2\pi}} \exp\left(-\frac{E^2}{2}\right). \tag{4.1.18}$$

The polynomials orthogonal with respect to a Gaussian function of this form are known as Hermite polynomials (e.g. Abramowitz and Stegun 1972), $He_n(E)$, $n = 0, 1, 2, \ldots$, of which the first few are:

$$\begin{aligned} He_0(E) &= 1 \\ He_1(E) &= E \\ He_2(E) &= E^2 - 1 \\ He_3(E) &= E^3 - 3E \end{aligned} \tag{4.1.19}$$

$$He_4(E) = E^4 - 6E^2 + 3$$
$$He_5(E) = E^5 - 10E^3 + 15E.$$

An expression for $He_n(E)$, valid for all values of n, is furnished by the Rodriguez formula

$$He_n(E) = (-1)^n \exp\left(\frac{E^2}{2}\right) \frac{d^n}{dE^n}\left[\exp\left(-\frac{E^2}{2}\right)\right] \tag{4.1.20}$$

(Abramowitz and Stegun 1972). Since the normalization integral is given by the expression $\int_{-\infty}^{\infty} He_k^2(E)p_0(E)dE = k!$, eqn(4.1.7) becomes

$$p(E) = \left(\frac{1}{2\pi}\right)^{1/2} \exp\left(-\frac{E^2}{2}\right) \sum_{k=0}^{\infty} \frac{\langle He_k(E)\rangle He_k(E)}{k!}. \tag{4.1.21}$$

In further applications we shall assume that all the atoms are located in general positions, and the components of the corresponding position vectors are rationally independent. Under these assumptions the pdf $p(E)$ is an even function of E, and hence all the odd moments of E vanish. We therefore obtain

$$p(E) = p_0(E) \sum_{k=0}^{\infty} \frac{\langle He_{2k}(E)\rangle He_{2k}(E)}{(2k)!} \tag{4.1.22}$$

$$= p_0(E) \left\{1 + \sum_{k=2}^{\infty} \frac{\langle He_{2k}(E)\rangle He_{2k}(E)}{(2k)!}\right\}. \tag{4.1.23}$$

The form of the expansion on the second line of eqn(4.1.23) is valid since the second moment of $|E|$ equals unity for all space-group symmetries and atomic compositions, which means that $\langle He_2(E)\rangle = 0$. This will be shown in some detail in section 4.2.

In the mathematical literature the function $p_0(E)$ is sometimes defined as

$$p_0(E) = \frac{1}{\sqrt{\pi}} \exp(-E^2), \tag{4.1.24}$$

in which case the orthogonal polynomials are denoted by $H_n(E)$. These are related to the $He_n(E)$ by $H_n(E) = 2^{n/2}He_n(E\sqrt{2})$. In order to avoid confusion in expansions in terms of Hermite polynomials the reader should always check which of the two versions of these polynomials is being used.

Similar expansions of the pdf may be derived for non-centrosymmetric crystals. Let us consider the form of the expansion for an acentric pdf based on the central limit theorem. In terms of the magnitude of the normalized structure factor, $|E|$, one finds as a lowest-order approximation $p_0(|E|) =$

$2|E|\exp(-|E|^2)$. Alternatively, one may work in terms of the normalized intensity $z = |E|^2$, the pdf of which is

$$p_0(z) \simeq e^{-z}, z \geq 0. \tag{4.1.25}$$

The polynomials orthogonal with respect to this pdf are known as Laguerre polynomials. These may conveniently be generated from the expression (Abramowitz and Stegun 1972):

$$L_n(z) = \frac{e^z}{n!}\frac{d^n}{dz^n}[z^n e^{-z}], n = 0, 1, 2, 3, \ldots . \tag{4.1.26}$$

The first few terms of an expansion of $p(z)$ in terms of $p_0(z)$ and Laguerre polynomials are explicitly

$$p(z) = e^{-z}\left\{1 + \frac{a_2}{2!}L_2(z) - \frac{a_3}{3!}L_3(z) + \frac{a_4}{4!}L_4(z) - \cdots\right\}, \tag{4.1.27}$$

where a_2 to a_4 are shown to be

$$\begin{aligned} a_2 &= \langle z^2\rangle - 2 \\ a_3 &= \langle z^3\rangle - 9\langle z^2\rangle + 12 \\ a_4 &= \langle z^4\rangle - 16\langle z^3\rangle + 72\langle z^2\rangle - 72 \end{aligned} \tag{4.1.28}$$

(Shmueli and Wilson 1981).

When suitably truncated, eqns (4.1.23) and (4.1.27) give the general forms of higher approximations to pdfs of the magnitude of the structure factor. It will be shown in section 4.2 that the moments of this magnitude generally depend on the atomic composition of the asymmetric unit and the space-group symmetry of the crystal. We will also discuss some methods of calculating these moments. Practical forms of these expansions will be dealt with in section 4.3. We may point out here that different versions of an expansion of a pdf in terms of Hermite polynomials are referred to in the statistical literature as Gram–Charlier or Edgeworth expansions (e.g. Cramér 1946; Davis 1983). Equations (4.1.21) and (4.1.27) are close to the Gram–Charlier form, even though one of them contains orthogonal polynomials other than the Hermite polynomials.

Other expansions in terms of the polynomials orthogonal to pdfs based on the central limit theorem include higher approximations to the ideal subcentric pdf (see section 3.4) and the ideal bicentric pdf (see section 3.3). These approximations will be discussed in section 4.4.

There is still another technique for deriving corrections to the lowest-order Gaussian approximation for pdfs for crystals composed of not too dissimilar atoms. More precisely, we will assume that the number of atoms in the unit cell is equal to N, the atomic numbers for each constituent atom being equal, and further assume in our development that N is large. A pdf of a (real) centrosymmetric structure factor will be considered. We have already derived an expansion of the pdf in one dimension based on this assumption in eqn(3.1.11). This can be converted to an expansion in terms of Hermite polynomials with the advantage that successive terms go in powers of $N^{-1/2}$. This allows an assessment of the order of the error incurred by truncating at a given term. Consider, for example, the expression for the characteristic function of the pdf for the sum of N random variables each with a mean equal to zero. When the characteristic function generated from the pdf of a single variable is equal to $C(\omega)$ then that for the sum of N equidistributed independent variables is equal to $C^N(\omega)$. Recall the definition of cumulants and the cumulant generating function given in eqn(3.1.14), and write

$$C^N(\omega) = \exp\left[N \sum_{n=2}^{\infty} \frac{i^n \kappa_n \omega^n}{n!}\right] \sim \exp\left(-N\kappa_2 \frac{\omega^2}{2}\right)\left[1 + N\frac{\kappa_4}{4!}\omega^4 + \cdots\right], \tag{4.1.29}$$

where only even cumulants have been retained, since in our applications $p(E)$ is an even function of E. It is advantageous, if one wants to perform a term-by-term Fourier inversion of the resulting series, to make use of the identity

$$\frac{1}{2\pi}\int_{-\infty}^{\infty} (i\omega)^n \exp[-i\omega S - \frac{\omega^2}{2}]d\omega = \frac{1}{\sqrt{2\pi}} \exp(-\frac{S^2}{2}) He_n(S). \tag{4.1.30}$$

We find that the pdf for the sum S is approximated by

$$q(S) \sim \frac{1}{\sqrt{2\pi N\kappa_2}} \exp\left(-\frac{S^2}{2N\kappa_2}\right)\left[1 + \frac{\kappa_4}{4!\kappa_2^{\frac{3}{2}} N^{\frac{3}{2}}} He_4\left(\frac{S}{\sqrt{N\kappa_2}}\right)\cdots\right]. \tag{4.1.31}$$

While such expansions are frequently encountered in the crystallographic literature the derivation of higher-order terms in the series of eqns (4.1.29) or (4.1.31) quickly turns into a tedious exercise. Nevertheless, techniques related to the Edgeworth expansion find their greatest use in multivariate problems, for which more exact methods become computationally difficult. However, when an extremely accurate approximation is a desideratum, e.g. if an improved approximation is to be constructed, one can use methods based on exact representations of the pdf in terms of Fourier series (Chapters 5 and 6). One can also use such techniques as exponentiation or the saddlepoint approximation (both of which are described below). All of these become computationally more difficult in higher dimensions, although no new ideas may be needed.

4.1.2 *Exponentiated pdfs*

The idea of exponentiation was first suggested by Charlier (1928) for general statistical applications and suggested for applications in crystallography by Karle (1972). The idea is a simple one which may be illustrated in the context of finding an approximation to the pdf of a univariate random variable. Let us return to the expansion in terms of eigenfunctions given in eqn(4.1.7) and rewrite the series as an exponential

$$c(x) \equiv \sum_{k=0}^{\infty} d_k \phi_k(x) = \exp\left[\sum_{n=1}^{\infty} b_n \phi_n(x)\right], \tag{4.1.32}$$

where the bs may be calculated in terms of the ds. The resulting series will be referred to as being an exponentiated one. Finding expressions for the b_n in terms of the d_k also requires rather complicated algebraic manipulations. Since the $\phi_k(x)$ are polynomials we may rewrite the series on the left-hand side of eqn(4.1.32) as

$$\sum_{k=0}^{\infty} d_k \phi_k(x) = \sum_{j=0}^{\infty} a_j x^j = \exp\left(\sum_{j=0}^{\infty} f_j x^j\right). \tag{4.1.33}$$

If one knows all of the coefficients of the polynomials, i.e. the d_k, then the a_js in this last equation are determined by expanding the polynomials into their component terms. When only a finite number of the ds are known, it is only possible to calculate approximate forms for the as. One next calculates the f_j from the series (Stuart and Ord, 1987) from the relation

$$f_j = \sum_{m=0}^{j} \sum a_1^{k_1} a_2^{k_2} \cdots a_m^{k_m} \frac{(-1)^{p-1}(p-1)!}{k_1! k_2! \cdots k_m!}, \tag{4.1.34}$$

in which the inner sum is to be taken over all non-negative integer values of the k_j and p subject to the conditions

$$k_1 + k_2 + \cdots + k_m = p \leq j. \tag{4.1.35}$$

The final step in the calculation is to expand x^j in terms of the $\phi_k(x)$ by writing

$$x^j = \sum_{k=0}^{j} A_{jk} \phi_k(x) \tag{4.1.36}$$

where, for example, A_{jk} is found by multiplying both sides of this equation by $p_0(x)$ and integrating over the entire relevant range of x. In this way one arrives at a formal expression for the b_n in eqn(4.1.32). Karle (1972) and

Karle and Gilardi (1973) applied a version of this technique to calculate an approximation to the joint pdf of three structure factors, claiming greater accuracy than obtainable using a straightforward expansion in orthogonal polynomials. At this time it is difficult to judge how general this improvement might be since no further applications of this technique appear to have been made.

4.1.3 *The saddlepoint approximation*

Most numerical evaluations of the pdfs required for the implementation of intensity statistics or direct methods are based on approximations furnished by the central limit theorem. Stripped of any probabilistic content, the central limit theorem can be regarded as a technique for the asymptotic evaluation of integrals containing a large parameter. In applications to crystallography the large parameter is generally related to the number of atoms in a unit cell. An example of this in a simplified setting is contained in our derivation of the Gaussian pdf of $E(\mathbf{h})$ in space group $P\bar{1}$ for a molecule with N identical atoms in the unit cell. This example was analysed in Chapter 2 and discussed more generally in section 3.1. That derivation was based on the assumption that $N \gg 1$ which permitted us to make the transition from an exact expression for the characteristic function in eqn(2.2.29) to an approximate one in eqn(2.2.30).

In spite of the widespread use of the central limit theorem approximation in crystallography it is deficient in a number of respects. The first and most obvious of these is that it can only furnish a Gaussian pdf in the lowest order of approximation. While the Gaussian pdf gives a qualitatively correct approximation when the atomic numbers are all more or less equal, it is unable to furnish a qualitatively correct picture when the atomic composition of the asymmetric unit is strongly heterogeneous. This, of course, is of practical importance in crystallographic applications. To remedy this deficiency, corrections based either on Edgeworth or Gram–Charlier expansions, as discussed in subsection 4.1.1, can be applied. However, such expansions generally converge slowly when atomic heterogeneity is pronounced. A slightly less evident shortcoming of the Gaussian approximation is that, even when it is qualitatively accurate, its accuracy is greatest in the neighbourhood of the peak, the error increasing often quite steeply towards the tails of the pdf.

In the remainder of this section we discuss another approach to finding approximate pdfs which we will refer to as the *saddlepoint approximation*, which is able to overcome at least some of the shortcomings of the central limit theorem. This advantage is obtained at the expense of requiring the solution of one or more transcendental equations.

The saddlepoint approximation is a method for finding an asymptotic approximation to integrals having the generic form

$$I_n = \int f(x)e^{-nx}dx \tag{4.1.37}$$

in one or more dimensions, in the limit $n \to \infty$ (de Bruijn, 1982). The first application of this general technique to a statistical problem was made by Cramér (1938) and it was later used in statistical physics by Khintchine (1949). It was then applied to a wider variety of problems in statistics and popularized in that field by Daniels in a series of papers (Daniels 1954, 1956, 1980, 1982). The saddlepoint approximation is not too well known in modern crystallographic applications, but Bricogne (1984) mentioned it as being closely related to the maximum entropy method. In this section we present a simple introduction to some of its properties and a discussion of some of its advantages particularly for applications characterized by substantial atomic heterogeneity.

Let us consider the simplest possible situation of a centrosymmetric crystal with N identical atoms in a unit cell of space group $P\bar{1}$ and let $C(\omega)$ be the characteristic function for a single atom. The pdf of the sum defining $E(\mathbf{h})$ will be denoted by $p(E)$. This function can be represented in terms of a Fourier transform as indicated in eqn(3.1.3). A first step in deriving an approximation is to rotate the axis along which the integral is to be evaluated through 90°. This is done by setting $\Omega = i\omega$, $\Gamma(\Omega) = C(-i\Omega)$, and $K(\Omega) = \ln \Gamma(\Omega)$. These functions allow us to express $p(E)$ as

$$p(E) = \frac{1}{2\pi i}\int_{-i\infty}^{i\infty} \Gamma^N(\Omega)e^{-\Omega E}d\Omega = \frac{1}{2\pi i}\int_{-i\infty}^{i\infty} \exp[N\{K(\Omega) - E\Omega/N\}]d\Omega. \tag{4.1.38}$$

The basic idea in the saddlepoint method coincides with that underlying the central limit theorem, being only a little more complicated in detail. Because of the factor N that multiplies the term in brackets in the exponent in eqn(4.1.38), the principal contribution to the value of the integral comes from the neighbourhood of the value of Ω at which the expression in brackets becomes an extremum. After finding this value, which we denote by Ω_0, one expands the integrand in a Taylor series in its neighbourhood, retaining the two lowest-order terms. The resulting integral can be evaluated exactly. Whereas in our derivation of the central limit theorem one knows that the maximum value of $C(\omega)$ occurs at $\omega = 0$, in the present instance the value of Ω_0 has generally to be calculated as the solution to one or more transcendental equations. Daniels has shown that only real values of Ω need be considered, and accordingly, that Ω_0 can be found as the root to

$$K'(\Omega) = x/N. \tag{4.1.39}$$

Daniels has further shown that in most circumstances this equation has only a single root (Daniels 1954).

We will see later, by example, that the saddlepoint method seems to work well not only when each of the constituent random variables in a sum has the same distribution, but also when the independent random variables have different distributions. This is significant in applications to crystallography because scattering factors for different types of atoms are different. Translated into more mathematical language this means that the components that contribute to the structure factor will have different pdfs when they come from different atomic species. If we say that there are a total of N atoms to be taken into account of which m_j are from species j, and the contribution from species j has a cumulant-generating function equal to $K_j(\Omega)$ then the function $K(\Omega)$ in eqn(4.1.38) is to be replaced by

$$K(\Omega) = \frac{m_1 K_1(\Omega) + m_2 K_2(\Omega) + \cdots + m_k K_k(\Omega)}{N} \tag{4.1.40}$$

in the formulae to follow. In this relation the parameter k is the total number of atomic species and the m_js are even integers as all the atoms are assumed to be located in general positions.

A derivation of the saddlepoint approximation to $p(E)$ is based on a change of the variable Ω to $\Omega_0 + iy$ in eqn(4.1.38). After this change of variables the next step is to expand the integrand in a series of powers of y. The lowest-order terms in this expansion are

$$K(\Omega) - \frac{x\Omega}{N} \sim K(\Omega_0) - \frac{x\Omega_0}{N} - \frac{1}{2}K''(\Omega_0)y^2 \tag{4.1.41}$$

where a term linear in y is missing because the derivative is equal to 0 as in eqn(4.1.39). The lowest-order saddlepoint approximation to $p(E)$ will be denoted by $p_{sp}(E)$, which is obtained by substituting eqn(4.1.41) into the integral in eqn(4.1.38), and integrating the result over all values of y. This yields an approximation which we write as

$$p_{sp}(E) \sim \frac{1}{\sqrt{2\pi N K''(\Omega_0)}} \exp\left[N\left\{K(\Omega_0) - \frac{E\Omega_0}{N}\right\}\right], \tag{4.1.42}$$

where $K(\Omega)$ is now given by eqn(4.1.40). The variable x is implicitly contained in the value of Ω_0 because of the defining relation in eqn(4.1.39). Higher-order corrections to the approximation in eqn(4.1.42) can be obtained by retaining further terms in the expansion of the left-hand side of eqn(4.1.41). Notice that, in contrast to the lowest-order Gaussian approximation resulting from the central limit theorem, $p_{sp}(E)$ is not necessarily properly normalized. In most applications one uses not this function directly but rather the normalized version

$$\overline{p}_{sp}(E) = \frac{p_{sp}(E)}{\int_{-\infty}^{\infty} p_{sp}(\xi)d\xi}. \tag{4.1.43}$$

Higher-order terms in the expansion of $p_{sp}(E)$, analogous to the Edgeworth series, are to be found in Daniels' original paper on the subject (1954). Another strategy for generating corrections is to express them in terms of the set of polynomials orthogonal to $p_{sp}(E)$.

A number of comments are in order at this point relating to the saddlepoint approximation in general. The first is that, in contrast to the formalism used in deriving the central limit theorem, the value of Ω_0 does not depend only on the local behaviour of $C(\omega)$ in the neighbourhood of the single point $\omega = 0$ but rather on the more general analytical behaviour of $C(\omega)$. Because of this, the function $p_{sp}(E)$ is generally not a Gaussian but may take a large variety of forms. The consequent flexibility permits one to derive approximations to pdfs that are not necessarily unimodal (as may be required when heavy atoms are present). The more significant property in the present discussion is that correction terms to the lowest approximation are generally smaller than those required for the Gaussian. Daniels showed in his 1954 paper on the subject that if one defines the quantities

$$\zeta_j = \frac{K^j(\Omega_0)}{[K''(\Omega_0)]^{\frac{j}{2}}}, \quad j = 1, 2, \cdots \tag{4.1.44}$$

then the next order of approximation is

$$p_{sp}^{(1)}(E) \sim p_{sp}(E)\left[1 + \frac{1}{N}\left(\frac{\zeta_4}{8} - \frac{5\zeta_3^2}{24}\right)\right], \tag{4.1.45}$$

which is to say that the error is of the order of $1/N$, as opposed to the $1/N^{1/2}$ found in the case when the variance is finite (Gnedenko and Kolmogorov 1954). Further terms in a series of corrections decrease as integer powers of N^{-1} rather than in powers of $N^{-1/2}$. The correction terms to the normalized pdf in eqn(4.1.43) exhibit even more favourable convergence properties (Daniels 1956), the first correction being of the order of $N^{-3/2}$ in place of the N^{-1} indicated in eqn(4.1.45). A final point is that the error in using the saddlepoint approximation is more uniformly distributed across the range spanned by the pdf than the error obtained resulting from the Gaussian approximation. Indeed, when the tail of a pdf contains useful information the saddlepoint approximation should be the technique of choice for approximating to the form of the pdf.

A simple calculation suffices to show that if the pdf for a single step is a Gaussian then the saddlepoint approximation leads to an exact result for the pdf for the sum of n such random variables. A more apposite example is provided by finding an approximation to the pdf of $E(\mathbf{h})$ for a crystal in space group $P\overline{1}$, which can be represented as shown in eqn(2.2.27). The characteristic function for such a crystal is

$$C(\omega) = \prod_{j=1}^{N/2} J_0(2n_j\omega) \tag{4.1.46}$$

so that $p(E)$ in this example is given by

$$p(E) = \frac{1}{2\pi i}\int_{-i\infty}^{i\infty} \exp(-E\Omega)\prod_{j=1}^{N/2} I_0(2n_j\Omega)d\Omega, \tag{4.1.47}$$

where $I_0(x)$ is a Bessel function. On taking the derivative of the product in eqn(4.1.47) we find that the equation that determines the root Ω_0 is

$$\sum_{j=1}^{N/2} n_j \frac{I_1(2n_j\Omega)}{I_0(2n_j\Omega)} = \frac{E}{2}. \tag{4.1.48}$$

Since the ratio $I_1(x)/I_0(x) < 1$ for any finite x it is obvious that this equation can have no solution unless E is less than $\sum_j n_j$ which confirms our earlier remark that the saddlepoint approximation to $p(E)$ is identically equal to zero when it is supposed to be, in contrast to the Gaussian approximation to the pdf.

With Ω defined by the solution to eqn(4.1.48) the unnormalized saddlepoint approximation to p(E) is

$$p_{sp}(E) = \frac{1}{\left[4\pi \sum_j n_j^2 \psi(2n_j E)\right]^{1/2}} e^{-\Omega E} \prod_{j=1}^{N/2} I_0(2n_j\Omega), \tag{4.1.49}$$

in which the function $\psi(x)$ is

$$\psi(x) = 1 - \frac{I_1(x)}{xI_0(x)} - \left[\frac{I_1(x)}{I_0(x)}\right]^2. \tag{4.1.50}$$

It can be shown that when

$$\frac{E^2}{4\sum_j n_j^2} \gg 1 \tag{4.1.51}$$

the approximation furnished by the saddlepoint method coincides with the Gaussian in the neighbourhood of $E = 0$ (Weiss and Kiefer 1983). When atomic heterogeneity is negligible then approximations furnished by both the central limit theorem and the saddlepoint method tend to give results which are quite accurate provided that the number of atoms in the unit cell is not too small. However, when atomic heterogeneity is important the saddlepoint approximation can provide a considerable improvement in accuracy relative to the Gaussian furnished by the central limit theorem.

To illustrate this last point we compare results for space group $P\bar{1}$ obtained by the Gaussian approximation corrected by an Edgeworth series, and the saddlepoint approximation using the same numbers of moments inserted into a correction series in terms of orthogonal polynomials. The criterion used in our comparison will be expressed as the integral

$$I_n = \int_0^{E_{max}} [p_{exact}(E) - p_{approx}(E)]^2 dE, \tag{4.1.52}$$

where n is the number of moments used in the calculation of the Edgeworth and saddlepoint approximate pdfs. Those are denoted by $p_{approx}(E)$. In eqn(4.1.52) $p_{exact}(E)$ is the exact Fourier representation of the pdf of E for space group $P\bar{1}$ (Weiss, Shmueli, Kiefer and Wilson 1985; see also section 5.1). The results of the comparison are given in Table 4.1 for unit cells containing a total of 30 atoms, and corrections calculated using up to five moments. The tabulated data indicate that with identical atoms both approximations give extremely accurate results, but when atomic heterogeneity is significant the accuracy of both approximations is noticeably degraded. In this particular example corrections to the saddlepoint approximation are considerably more effective in increasing the accuracy of the approximation than an Edgeworth series is in correcting the Gaussian approximation.

The saddlepoint approximation can be extended to higher dimensions (Good, 1961), but has not yet been applied to calculate pdfs required for the implementation of direct methods. It would be interesting to see whether the increased amount of computing required to solve the basic saddlepoint equation is excessive in applications having a considerable degree of atomic heterogeneity.

4.2 Moments of the structure factor

We have described different methods for constructing higher approximations to the pdf in section 4.1. These generally improve the accuracy of the theoretical pdfs for the normalized structure factor or its magnitude over the result derived from the Gaussian or non-Gaussian pdf given by the central limit theorem. Such lowest-order approximations to exact pdfs

Table 4.1 A comparison of discrepancies given by eqn(4.1.52) obtained using an Edgeworth series based on the Gaussian approximation, and the saddlepoint approximation together with an expansion in terms of polynomials orthogonal to the lowest-order term. The asymmetric half of the unit cell consists of 15 atoms. In case 1 we have taken $n_1 = n_2 = \cdots = n_{15} = 1$, and case 2 is defined by $n_1 = n_2 = n_{14} = 1$, $n_{15} = 15\frac{1}{3}$ which corresponds to having 14 carbon and a single uranium atom in the asymmetric unit of space group $P\overline{1}$.

	Case 1		Case 2	
	Homogeneous unit cell		Heterogeneous unit cell	
n	Edgeworth	Saddlepoint	Edgeworth	Saddlepoint
0	1.2 (E-05)	4.8 (E-07)	3.5 (E-01)	5.6 (E-02)
1	1.2 (E-05)	3.5 (E-09)	3.5 (E-01)	1.4 (E-02)
2	9.5 (E-09)	9.7 (E-12)	3.2 (E-01)	2.5 (E-03)
3	6.4 (E-08)	2.0 (E-11)	1.8 (E-01)	2.8 (E-03)
4	1.3 (E-09)	1.5 (E-11)	1.1 (E-01)	1.2 (E-03)
5	8.5 (E-10)	6.6 (E-12)	7.9 (E-02)	3.4 (E-04)

are sometimes termed ideal pdfs. The most straightforward of these alternative methods and the only one so far applied to arbitrary atomic composition and all space-group symmetries is the one described in subsection 4.1.1. That method consists of approximating the exact pdf by a product of the ideal pdf and a correcting expansion generated in terms of orthogonal polynomials. The weight function used to generate these polynomials is obtained from the central limit theorem. For example, the ideal centric and acentric pdfs, given by eqns (3.2.49) and (3.2.58), are weight functions for Hermite and Laguerre orthogonal polynomials given by eqns (4.1.20) and (4.1.26) respectively. The coefficients of the correcting expansion are linear combinations of the moments of the exact pdf [see, for example, eqns (4.1.7) and (4.1.23)], hence a knowledge of these moments is a prerequisite for the construction of any such approximation. Furthermore, as mentioned earlier, these moments can be computed exactly without knowing the exact pdf. In this section we shall discuss the problem of calculating moments of the pdfs of the structure factor knowing the functional form of the corresponding trigonometric structure factor and the atomic composition of the crystal. Our presentation is based on the works of Wilson (1978), Shmueli and Wilson (1981, 1982), Shmueli and Kaldor (1981,1983) and Shmueli

(1982). However, we confine ourselves to detailed examples of moments of low orders, and refer the reader to more comprehensive results presented in these and other references.

In the calculations to follow we shall make the following assumptions:

1. Atomic contributions to the structure factor are statistically independent.
2. The components of the atomic position vectors are rationally independent so that the argument of the atomic phase factor is uniformly distributed over $(-\pi, \pi)$.
3. Dispersion is negligible.
4. There is no non-crystallographic symmetry.

We begin by reformulating the normalized structure factor in terms of the atomic composition of the asymmetric unit and space-group symmetry. We have

$$\begin{aligned} E(\mathbf{h}) &= \sum_{j=1}^{N} n_j \exp(2\pi i \mathbf{h}^T \mathbf{r}_j) \\ &= \sum_{j=1}^{N/g} n_j \sum_{s=1}^{g} \exp[2\pi i \mathbf{h}^T (\mathbf{P}_s \mathbf{r}_j + \mathbf{t}_s)] \\ &\equiv \sum_{j=1}^{N/g} n_j T_j(\mathbf{h}), \end{aligned} \tag{4.2.53}$$

where n_j is the normalized scattering factor and

$$T_j = \sum_{s=1}^{g} \exp[2\pi i \mathbf{h}^T (\mathbf{P}_s \mathbf{r}_j + \mathbf{t}_s)] \tag{4.2.54}$$

is the (complex) trigonometric structure factor of the jth atom. In this expression g is the order of the point group of the crystal times the multiplicity of its Bravais lattice and $(\mathbf{P}_s, \mathbf{t}_s)$ is the sth operator of the space group which transforms the position vector of the jth atom to that of an equivalent one.

It is instructive to consider in detail the general calculation of the second moment of $|E|$. We have

$$\begin{aligned} \langle |E|^2 \rangle &= \langle EE^* \rangle \\ &= \sum_{j=1}^{N/g} \sum_{k=1}^{N/g} n_j n_k \langle T_j T_k^* \rangle, \end{aligned} \tag{4.2.55}$$

since dispersion has been neglected. The inner average is given by

$$\langle T_j T_k^* \rangle = \sum_{s=1}^{g} \sum_{u=1}^{g} \langle \exp[2\pi i \mathbf{h}^T(\mathbf{P}_s \mathbf{r}_j - \mathbf{P}_u \mathbf{r}_k)] \rangle \exp[2\pi i \mathbf{h}^T(\mathbf{t}_s - \mathbf{t}_u)]. \quad (4.2.56)$$

Since the quadratic forms $\mathbf{h}^T \mathbf{P}_s \mathbf{r}_j$ and $\mathbf{h}^T \mathbf{P}_u \mathbf{r}_k$ are independent and each is uniformly distributed over [0,1], the inner average vanishes whenever $j \neq k$. We can therefore rewrite eqn(4.2.56) as

$$\begin{aligned} \langle |E|^2 \rangle &= \sum_{j=1}^{N/g} n_j^2 \langle T_j T_j^* \rangle \\ &= \sum_{j=1}^{N/g} n_j^2 \langle |T_j|^2 \rangle. \end{aligned} \quad (4.2.57)$$

Further, we can write

$$\langle |T_j|^2 \rangle = \sum_{s=1}^{g} \sum_{u=1}^{g} \langle \exp[2\pi i \mathbf{h}^T(\mathbf{P}_s - \mathbf{P}_u)\mathbf{r}_j)] \rangle \exp[2\pi i \mathbf{h}^T(\mathbf{t}_s - \mathbf{t}_u)]. \quad (4.2.58)$$

Again, the inner average vanishes unless $\mathbf{P}_s - \mathbf{P}_u$ is a zero matrix, in which case it equals unity. Hence it follows that

$$\langle |T_j|^2 \rangle = \sum_{s=1}^{g} \sum_{u=1}^{g} \delta_{su} \exp[2\pi i \mathbf{h}^T(\mathbf{t}_s - \mathbf{t}_u)] = g. \quad (4.2.59)$$

Finally, the expression for $\langle |E|^2 \rangle$ is

$$\langle |E|^2 \rangle = g \sum_{j=1}^{N/g} n_j^2 = \sum_{j=1}^{N} n_j^2 = 1, \quad (4.2.60)$$

which is, of course, familiar to anyone involved in structure determination. Note, however, that the result is valid for arbitrary atomic heterogeneity as well as for any space-group symmetry. Its derivation will also be useful in the discussion to follow.

4.2.1 *Examples*

Before proceeding to the next example, it is convenient to summarize some statistical properties of products of trigonometric structure factors (Wilson 1978). Firstly, we note that

$$\langle T_j T_k^* \rangle = \begin{cases} 0 & \text{if } j \neq k \\ \langle |T_j|^2 \rangle & \text{if } j = k \end{cases} \quad (4.2.61)$$

for centrosymmetric as well as non-centrosymmetric space groups. A second identity is

$$\langle T_j T_j \rangle = \begin{cases} 0 & \text{if the crystal is non-centrosymmetric} \\ \langle |T_j|^2 \rangle & \text{if the crystal is centrosymmetric} \end{cases} \tag{4.2.62}$$

since for the centrosymmetric case $T_j = T_j^*$, and finally

$$\langle T_j T_k \rangle = 0 \text{ if } j \neq k, \text{ for any space group.} \tag{4.2.63}$$

We can now proceed to a somewhat less trivial example, the calculation of the fourth moment of $|E|$. We have

$$\langle |E|^4 \rangle = \sum_{i=1}^{N/g} \sum_{j=1}^{N/g} \sum_{k=1}^{N/g} \sum_{l=1}^{N/g} n_i n_j n_k n_l \langle T_i T_j^* T_k T_l^* \rangle. \tag{4.2.64}$$

In view of eqns (4.2.61), (4.2.62) and (4.2.63), the surviving inner averages in eqn(4.2.64) must have i, j, k and l equal in pairs. The two allowed combinations for non-centrosymmetric space groups are $i = j$ and $k = l$ or $i = l$ and $j = k$, while three combinations are allowed for centrosymmetric space groups: $i = j$ and $k = l$ or $i = l$ and $j = k$ or $i = k$ and $j = l$. We therefore have

$$\langle |E|^4 \rangle = L \sum_{i \neq k} n_i^2 n_k^2 \langle |T_i|^2 \rangle \langle |T_k|^2 \rangle + \sum_j n_j^4 \langle |T_j|^4 \rangle, \tag{4.2.65}$$

where L equals 2 or 3 according as the crystal is non-centrosymmetric or centrosymmetric respectively. If we make use of the identity

$$\sum_{i \neq k} x_i x_k \equiv \left(\sum_i x_i \right)^2 - \sum_i x_i^2 \tag{4.2.66}$$

eqn(4.2.65) reduces to

$$\langle |E|^4 \rangle = L \left(\sum_i^{N/g} n_i^2 \langle |T_i|^2 \right)^2 + \sum_{j=1}^{N/g} (\langle |T_j|^4 \rangle - L \langle |T_j|^2 \rangle^2) n_j^4 \tag{4.2.67}$$

$$= L \langle |T|^2 \rangle^2 \left(\sum_{i=1}^{N/g} n_i^2 \right)^2 + (\langle |T|^4 \rangle - L \langle |T|^2 \rangle^2) \sum_j^{N/g} n_j^4. \tag{4.2.68}$$

The atomic subscripts have been dropped from T since all the atoms are assumed to be located in the same (general) Wyckoff position. We recall that $\langle|T|\rangle \sum_{i=1}^{N/g} n_i^2 = \langle|E|^2\rangle = 1$, and divide both sides of eqn(4.2.68) by the square of this quantity. In this way we finally obtain

$$\langle|E|^4\rangle = L + \left(\frac{\langle|T|^4\rangle}{\langle|T|^2\rangle^2} - L\right) Q_4, \tag{4.2.69}$$

where

$$Q_4 = \frac{\sum_{j=1}^{N/g} n_j^4}{\left(\sum_{i=1}^{N/g} n_i^2\right)^2} = \frac{\sum_{j=1}^{N/g} f_j^4}{\left(\sum_{i=1}^{N/g} f_i^2\right)^2} \simeq \frac{\sum_{j=1}^{N/g} Z_j^4}{\left(\sum_{i=1}^{N/g} Z_i^2\right)^2}. \tag{4.2.70}$$

A derivation of the fourth moment of $|F|$, taking dispersion into account has been given by Wilson (1978).

It remains to find a space-group dependent expression for $\langle|T|^4\rangle$. This can be found for most space groups by a method analogous to that employed in the evaluation of $\langle|T|^2\rangle$. We have

$$\langle|T|^4\rangle = \sum_{s=1}^{g}\sum_{t=1}^{g}\sum_{u=1}^{g}\sum_{v=1}^{g} \langle\exp(i\theta_{stuv})\rangle \exp(i\tau_{stuv}), \tag{4.2.71}$$

where

$$\theta_{stuv} = 2\pi\mathbf{h}^T(\mathbf{P}_s - \mathbf{P}_t + \mathbf{P}_u - \mathbf{P}_v)\mathbf{r} \tag{4.2.72}$$

and

$$\tau_{stuv} = 2\pi\mathbf{h}^T(\mathbf{t}_s - \mathbf{t}_t + \mathbf{t}_u - \mathbf{t}_v). \tag{4.2.73}$$

The assumptions made earlier imply that the inner average in eqn(4.2.71) vanishes unless $\mathbf{P}_s - \mathbf{P}_t + \mathbf{P}_u - \mathbf{P}_v$ is a zero matrix. A simple computer algorithm can be constructed, by which the moments of $|T|$ can be rapidly computed for all space groups with $g \leq 24$. The effective flow chart for computing the fourth moment of $|T|$ consists of three steps, namely:

1. Read the space-group operators $(\mathbf{P}_s, \mathbf{t}_s)$ as matrix–vector pairs, or reconstruct them in the program from the symbolic coordinates of equivalent general positions as given in *International Tables* (1983).
2. Open a fourfold loop on the summation indices s, t, u, v, while making full use of the symmetry of the summation. Storage of differences between space-group operators simplifies coding.
3. For each combination of s, t, u, v for which the four-matrix sum in θ_{stuv} is a zero matrix, accumulate values of $\exp(i\tau_{stuv})$.

The result is a numerical value of the fourth moment of $|T|$.

Expressions for even moments of $|T|$ can be readily obtained in closed form for all the space groups of the triclinic, monoclinic and orthorhombic systems except for $Fdd2$ and $Fddd$ (Shmueli 1982). For example, let us consider a calculation of the even moments of $|T|$ for $P\bar{1}$. The trigonometric structure factor is here given by $T_j = 2\cos\theta_j$. Hence

$$\begin{aligned}\langle |T|^{2n}\rangle &= 2^{2n}\langle \cos^{2n}\theta\rangle \\ &= 2^{2n}\left[\frac{1}{2\pi}\int_{-\pi}^{\pi}\cos^{2n}\theta d\theta\right] \qquad (4.2.74)\\ &= \frac{(2n)!}{(n!)^2}. \qquad (4.2.75)\end{aligned}$$

The normalized even moment of $|T|$ for this space group is given by

$$\gamma_{2n} = \frac{\langle |T|^{2n}\rangle}{\langle |T|^2\rangle^n} = \frac{(2n)!}{2^n(n!)^2}. \qquad (4.2.76)$$

Shmueli and Kaldor (1981,1983) showed that the moments of $|T|$ for these low symmetries depend on the point-group rather than on the space-group symmetry. Closed expressions for the normalized moments of $|T|$ for the triclinic, monoclinic and orthorhombic space groups, not including those for $Fdd2$ and $Fddd$, are given in Table 4.2.

Table 4.2 The normalized even moments of $|T|$. These are expressed in terms of M_n, where

$$M_n = \frac{(2n)!}{2^n(n!)^2}$$

and l, which takes the values 1, 2 or 4 according as the Bravais lattice is of type P, one of the types A, B, C or I, or type F respectively. The expressions for γ_{2n} (see text) are identical for all the space groups isomorphic to a given point group with the two exceptions mentioned above. The expressions are valid for general reflections and subject to assumptions given in the text.

Point group(s)	Expression for γ_{2n}
1	1
$\bar{1}$, 2, m	$l^{n-1}M_n$
$2/m$, $mm2$	$l^{n-1}M_n^2$
mmm	$l^{n-1}M_n^3$
222	$\frac{l^{n-1}}{2^n}\sum_{p=0}^{n}(M_{n-p}M_p)^3\binom{k}{p}^{-2}$

The fourth, sixth and eighth moments of the magnitude of the trigonometric structure factor are given for the 230 space groups, and all parities of the reflection indices for which the moments assume different values, by Shmueli and Kaldor (1981,1983). Moments for space groups with $g > 24$ were computed with the aid of an algorithm which was based on a computer–algebraic treatment of the simplified forms of the real and imaginary parts of the trigonometric structure factor.

4.2.2 *Effects of space-group symmetry and atomic heterogeneity on the fourth moment of* $|E|$

Shmueli and Wilson (1981) have examined the effects of space-group symmetry and the presence of outstandingly heavy atoms on the values of the even moments of $|E|$, by taking the fourth moment as an example. Suppose that the asymmetric unit of the space group consists of m equal light scatterers of atomic number Z_L and a single heavy scatterer of atomic number Z_H. The composition-dependent part of the moment [see eqns (4.2.69) and (4.2.70)] can now be written as

$$Q_4 = \frac{mZ_L^4 + Z_H^4}{(mZ_L^2 + Z_H^2)^2} = \frac{m + \rho^4}{(m + \rho^2)^2}, \qquad (4.2.77)$$

where $\rho = Z_H/Z_L$. We see from the combination of eqn(4.2.69) and this equation that in the equal-atom case, i.e. for $\rho = 1$, and m sufficiently large, the fourth moment of $|E|$ can be approximated by 2 or 3 according as the space group is non-centrosymmetric or centrosymmetric respectively. As shown in section 3.2 these are the values predicted by the central limit theorem. At the other extreme, with a considerable degree of atomic heterogeneity, i.e. for $\rho \gg 1$, Q_4 tends to unity and $\langle |E|^4 \rangle$ tends to γ_4, i.e. it is determined by the symmetry alone. The dependence of $\langle |E|^4 \rangle$ on ρ, for all centrosymmetric and non-centrosymmetric symmorphic space groups with Bravais lattices of type P, is illustrated in Figs 4.2 and 4.3. We see that the departure of the fourth moment of $|E|$ from the value (2 or 3) predicted by the central limit theorem generally increases as the degree of heterogeneity increases. However, this increase may depend significantly on symmetry. For some space groups the departure appears to be so significant that predictions by the central limit theorem are somewhat or entirely misleading (e.g. the moment for $P6mm$ appears to tend to the ideal centric value and that for $P\overline{1}$ can be mistaken for one arising from an ideal acentric distribution). For others the value of the moment is practically insensitive to the presence of the heavy atom in the asymmetric unit. Besides the wide range of behaviour, the curves also display a 'heterogeneity saturation'. In fact, Figs 4.2 and 4.3 show that for strong heterogeneity the value of the fourth moment remains approximately constant as the degree of heterogeneity in-

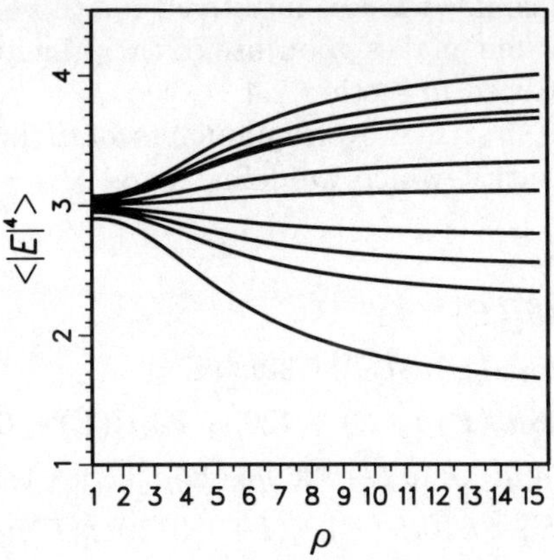

FIG. 4.2. *The exact fourth moment of $|E|$ as a function of atomic heterogeneity for some centrosymmetric space groups.*

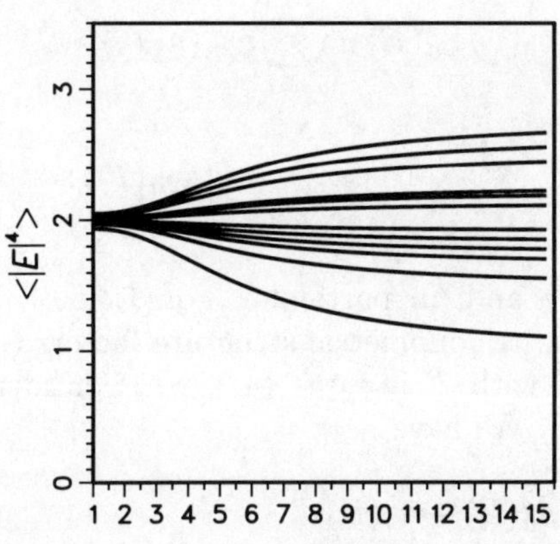

FIG. 4.3. *The exact fourth moment of $|E|$ as a function of atomic heterogeneity for some non-centrosymmetric space groups.*

creases. This is of course expected on the basis of our earlier qualitative discussion of extreme compositions.

4.2.3 *Moments of the structure factor from moment–cumulant relationships*

An alternative method of finding the relationship between the moments of the structure factor and moments of the trigonometric structure factor was proposed by Shmueli and Wilson (1982) for the centrosymmetric case. The method is somewhat less intuitive than those discussed in earlier parts of this section, but makes good use of the principles of the theory of probability as introduced in section 1.4.

Let us write the first five even cumulants of E in terms of the even moments of this quantity, which we define as $\mu_{2n}(E) = \langle |E|^{2n} \rangle$.

$$\begin{aligned}
\kappa_2(E) &= \mu_2(E) \\
\kappa_4(E) &= \mu_4(E) - 3\mu_2^2(E) \\
\kappa_6(E) &= \mu_6(E) - 15\mu_4(E)\mu_2(E) + 30\mu_2^3(E) \qquad (4.2.78)\\
\kappa_8(E) &= \mu_8(E) - 28\mu_6(E)\mu_2(E) + 420\mu_4(E)\mu_2^2(E) - 630\mu_2^4(E) - 35\mu_4^2(E) \\
\kappa_{10}(E) &= \mu_{10}(E) - 45\mu_8(E)\mu_2 + 1260\mu_6(E)\mu_2^2(E) - 18900\mu_4(E)\mu_2^3(E) \\
&\quad + 22680\mu_2^5(E) - 210\mu_6(E)\mu_4(E) + 3150\mu_4^2(E)\mu_2(E).
\end{aligned}$$

In view of the additivity of cumulants of independent random variables, and their transformation properties, the even cumulants of E can be expressed in terms of those of the trigonometric structure factor as

$$\begin{aligned}
\kappa_{2n}(E) &= \sum_{j=1}^{N/g} \kappa_{2n}(n_j T_j) \\
&= \sum_{j=1}^{N/g} n_j^{2n} \kappa_{2n}(T_j) \qquad (4.2.79)
\end{aligned}$$

[see subsection 1.4.8 and, in particular, eqn(1.4.60)]. We can now relate the cumulants of the trigonometric structure factors to their moments by simply replacing E with T in eqns (4.2.78). Consider, for example, the fourth moment of E. We have

$$\begin{aligned}
\mu_4(E) &= 3\mu_2^2(E) + \kappa_4(E) \qquad (4.2.80)\\
&= 3\kappa_2^2(E) + \kappa_4(E) \qquad (4.2.81)\\
&= 3\left(\sum_{i=1}^{N/g} n_i^2 \mu_2(T_i)\right)^2 + \sum_{j=1}^{N/g} n_j^4[\mu_4(T_j) - 3\mu_2^2(T_j)], \qquad (4.2.82)
\end{aligned}$$

where $\mu_{2n}(T_j) = \langle |T_j|^{2n} \rangle$. We drop the atomic subscripts from the Ts and recall that $\mu_2(T)\sum_{j=1}^{N/g} n_j^2 = \mu_2(E) = 1$. If we now divide eqn(4.2.81) by $\mu_2^2(E)$, we obtain

$$\mu_4(E) = 3 + \left(\frac{\mu_4(T)}{\mu_2^2(T)} - 3\right) \frac{\sum_{j=1}^{N/g} n_j^4}{\left(\sum_{i=1}^{N/g} n_i^2\right)^2} \qquad (4.2.83)$$

in agreement with eqn(4.2.69). Expressions for higher moments of E can be obtained along similar lines.

4.3 Some approximate non-ideal distributions

The previous section dealt with some methods of deriving the even moments of $|E|$, needed for expanding the pdf in terms of orthogonal polynomials as in eqn(4.1.7), and with the dependence of these moments on the absolute moments of the trigonometric structure factor, T. As pointed out earlier the fourth, sixth and eighth moments of $|T|$ were evaluated for all the space groups and parity groups of reflection indices for which T has different functional forms for the same space group. Moreover, closed expressions for even moments of $|T|$, of any order, are available for all triclinic, monoclinic and orthorhombic space groups with the exception of $Fdd2$ and $Fddd$.

4.3.1 *Dispersionless centric and acentric pdfs*

We can now summarize generalized centric and acentric pdfs of $|E|$, having the form of the expansion in orthogonal polynomials on the right-hand side of eqn (4.1.7), and subject to the assumptions made at the beginning of the last section. In particular, we neglect effects of dispersion and non-crystallographic symmetry which, however, will be dealt with in subsection 4.3.2. Our summary is based on the works of Shmueli and Wilson (1981) and Shmueli (1982). The appropriate pdfs based on the central limit theorem, as given by Wilson (1949) statistics and denoted here by $p^o(|E|)$, and the corresponding orthogonal polynomials, $\phi_k(|E|)$, are shown in Table 4.3,

Table 4.3

$p^o(\lvert E\rvert)$	$\phi_k(\lvert E\rvert)$	Distribution
$(2/\pi)^{1/2}\exp(-\lvert E\rvert^2/2)$	$He_{2k}(\lvert E\rvert)/[(2k)!]^{1/2}$	Centric
$2\lvert E\rvert\exp(-\lvert E\rvert^2)$	$L_k(\lvert E\rvert^2)$	Acentric

where $He_k(x)$ and $L_k(x)$ are Hermite and Laguerre polynomials respectively (e.g. Abramowitz and Stegun 1972). The resulting pdfs of $|E|$ are given by

$$p_c(|E|) = p_c^o(|E|)\left\{1 + \sum_{k=2}^{\infty}\frac{A_{2k}}{(2k)!}He_{2k}(|E|)\right\} \tag{4.3.84}$$

and

$$p_a(|E|) = p_a^o(|E|)\left\{1 + \sum_{k=2}^{\infty}\frac{(-1)^k B_{2k}}{k!}L_k(|E|^2)\right\} \tag{4.3.85}$$

for centrosymmetric and non-centrosymmetric space groups respectively, where $p_c^o(|E|)$ and $p_a^o(|E|)$ are the centric and acentric pdfs based on the central limit theorem respectively (see Table 4.3). The expansion coefficients A_{2k} and B_{2k} can be related to the even moments of $|E|$ by evaluating the integrals $\langle|E|^{2k}\rangle = \int_0^\infty E^{2k}p_\alpha(|E|)d|E|$, where p_α is given by eqn(4.3.84) and eqn(4.3.85) respectively [see eqns (A.0.6) and (A.0.7) in Appendix A]. The result of these calculations is

$$\langle|E|^{2k}\rangle = \alpha_{k0} + \begin{cases}\sum_{p=2}^{k}\alpha_{kp}A_{2p} \text{ if centric}\\ \sum_{p=2}^{k}\alpha_{kp}B_{2p} \text{ if acentric}\end{cases} \tag{4.3.86}$$

with

$$\alpha_{kp} = \binom{k}{p}\frac{\langle|E|^{2k}\rangle_0}{\langle|E|^{2p}\rangle_0}, \tag{4.3.87}$$

where $\langle|E|^{2k}\rangle_0 = (2k)!/(2^k k!)$ or $k!$ according as the space group is centrosymmetric or non-centrosymmetric respectively. The unified forms of the coefficients A_{2k} and B_{2k}, so far available for k =2, 3, 4 and 5, are

$$A_4 \text{ or } B_4 = a_4Q_4 \tag{4.3.88}$$
$$A_6 \text{ or } B_6 = a_6Q_6 \tag{4.3.89}$$
$$A_8 \text{ or } B_8 = a_8Q_8 + U(a_4^2Q_4^2 - \gamma_4^2Q_8) \tag{4.3.90}$$

and

$$A_{10} \text{ or } B_{10} = a_{10}Q_{10} + V(a_4a_6Q_4Q_6 - \gamma_4\gamma_6Q_{10}) + W\gamma_4^2Q_{10} \tag{4.3.91}$$

(Shmueli 1982), where U = 35 or 18, V = 210 or 100 and W = 3150 or 900 according as A_{2k} or B_{2k} is required respectively; the other quantities in eqns (4.3.88)–(4.3.91) are given below. The *composition-dependent* terms in these equations are

$$Q_{2k} = \frac{\sum_{j=1}^m f_j^{2k}}{\left(\sum_{n=1}^m f_n^2\right)^k}, \tag{4.3.92}$$

where m is the number of atoms in the asymmetric unit and $f_j, j = 1\ldots m$, are their scattering factors, and the *dependence on symmetry* is expressed by the coefficients a_{2k} in eqns(4.3.88)–(4.3.91) as follows:

$$a_{2k} = (-1)^{k-1}(k-1)!\alpha_{k0} + \sum_{p=2}^{k}(-1)^{k-p}(k-p)!\alpha_{kp}\gamma_{2p}, \tag{4.3.93}$$

where α_{kp} is defined in eqn(4.3.87) and γ_{2p} which appears in eqns (4.3.90), (4.3.91) and (4.3.93) is given by

$$\gamma_{2p} = \frac{\langle |T|^{2p}\rangle}{\langle |T|^2\rangle^p}. \tag{4.3.94}$$

In this equation $\langle |T|^{2p}\rangle$ are the even absolute moments of the trigonometric structure factor that have been described in section 4.2. As pointed out earlier, these moments have been calculated for all the space groups for $p = 2$, 3 and 4 only, and can be computed for any value of p for space groups lower than tetragonal, except $Fdd2$ and $Fddd$ (see Table 4.2). The so far available list of expansion coefficients A_{2k} and B_{2k} can be readily extended through the use of computerized symbolic programming methods.

We include, for the sake of completeness, the correponding expressions for the corrected cumulative distribution function of $|E|$, i.e. the probability that the magnitude of the normalized structure factor does not exceed $|E|$. These functions are

$$N_c(|E|) = \mathrm{erf}\left(\frac{|E|}{\sqrt{2}}\right) - \sqrt{\frac{2}{\pi}}\exp\left(-\frac{E^2}{2}\right) \times \left\{\sum_{k=2}^{\infty}\frac{A_{2k}}{(2k)!}He_{2k-1}(|E|)\right\} \tag{4.3.95}$$

and

$$N_a(|E|) = 1 - \exp(-|E|^2) + \exp(-|E|^2) \times \left\{\sum_{k=2}^{\infty}\frac{(-1)^k B_{2k}}{k!}[L_{k-1}(|E|^2) - L_k(|E|^2)]\right\}. \tag{4.3.96}$$

However, because of the inadequate perfomance of these approximations in the presence of outstandingly heavy scatterers in crystals belonging to some space groups, further developments of this correction-factor approach should be replaced by the more accurate Fourier methods that will be described in Chapter 5. Nevertheless, four-term expansions of the type described in this section appear to be sufficient for crystals belonging to the cubic space groups, even though they may be characterized by a very strong heterogeneity of atomic composition (Shmueli, Stein and Weiss 1993).

4.3.2 *Pdfs allowing for dispersion*

The approximations described in the previous subsection are rather straightforward once the moments of the trigonometric structure factor are available. This is so because the ideal pdfs discussed there are weight functions of well-known orthogonal polynomials. The situation appears to be more complicated when we consider ideal pdfs for which the orthogonal polynomials must be constructed *ab initio*. We shall consider here the calculation of corrections to the ideal subcentric pdf, with particular application to the problem of dispersion. The treatment of Shmueli and Wilson (1983) will be followed closely.

The ideal pdf of $z = |E|^2$, allowing for dispersion, is given by

$$p(z) = \alpha^{1/2} \exp(-\alpha z) I_0(\beta z), \tag{4.3.97}$$

where

$$\alpha = 1/(1 - S_E^2), \ \beta = \alpha S_E$$

and

$$S_E = \frac{\left|\sum_{j=1}^{N} (f_j' + i f_j'')^2\right|}{\sum_{j=1}^{N} |f_j' + i f_j''|^2}$$

(see section 3.4). In the last equation f_j' and f_j'' are the real and imaginary parts of the atomic scattering factor of atom j respectively. Following our discussion in subsection 4.1.1 the orthogonal polynomials whose weight function is eqn(4.3.97) can be constructed from Szegö determinants [see eqn(4.1.16)], for which the moments of $p(z)$ are required. As shown in section 3.4, these are given by

$$c_n = \alpha^{1/2} \int_0^\infty z^n \exp(-\alpha z) I_0(\beta z) dz \tag{4.3.98}$$

$$= n! \alpha^{-n/2} P_n(\alpha^{1/2}) \tag{4.3.99}$$

and if the Rodriguez formula for the Legendre polynomials, $P_n(x)$, is employed, we obtain for the moments of the ideal pdf (4.3.97)

$$c_n = (2x)^n \frac{d^n}{dx^n} (x^2 - 1)^n, \tag{4.3.100}$$

where $x = \alpha^{1/2}$. Equation (4.3.100) is then used for the generation of the moments required for the generation of the orthogonal polynomials from the determinants in eqn(4.1.16).

It may be of interest to note that the apparently cumbersome algebra involved in the above steps can be carried out in an error-free manner by computerized symbolic programming. For example, the calculation of the first eight moments, c_n, in terms of $x = \alpha^{1/2}$ requires the following line of REDUCE code

```
FOR K:=1:8 DO C(K):=(2*X)**(-K)*DF((X**2-1)**K, X, K);
```

where '1:8' means '1 until 8 with increment 1' and DF(F,X,K) calculates the Kth (partial) derivative of F with respect to X. The results, C(K), can be produced in symbolic form, or their output can be produced in a Fortran-compatible format which permits their insertion into a numeric program for further use. However, in the present application they can be used directly in further REDUCE statements, which calculate the orthogonal polynomials in symbolic form by the use of appropriate determinant expansion statements.

The next step is the calculation of moments of the exact pdf of z taking into account dispersion. This is appreciably more complicated than in the dispersionless case since the scattering factors are complex quantities. We shall content ourselves with a schematic description of this calculation, since an easier and accurate derivation of the pdf for dispersion will be given in section 5.8. The $2n$th absolute moment of $|E|$ (or the nth moment of the normalized intensity z) is given by

$$\begin{aligned} \langle |E|^{2n} \rangle &= \langle (EE^*)^n \rangle \\ &\equiv \Sigma^{2n} C\langle S \rangle, \end{aligned} \tag{4.3.101}$$

where the symbol Σ^{2n} denotes a $2n$-fold summation which involves indices ranging over the atoms of the asymmetric unit, C is a $2n$-fold product of atomic scattering factors (and their complex conjugates) and S is a $2n$-fold product of corresponding trigonometric structure factors. Thus, C depends on the composition of the crystal and on the extent of effects of dispersion, while S reflects the crystal symmetry. Since, however, the statistical properties of the structure factor and its moments are related to the symmetry terms S, the actual averaging is performed on these quantities and the composition terms C in eqn(4.3.101) are constant in this process.

As will be seen below, in the centrosymmetric case with the scattering factors being possibly complex, the composition term C can assume several forms for the same representation of $\langle S \rangle$, and each of these forms may have a different multiplicity. When the multiplicities of the $C\langle S \rangle$ arrangements are determined, the summation in eqn(4.3.101) can be decomposed into partial summations as

$$\langle |E|^{2n} \rangle = \sum_k m_k \Sigma' (C\langle S \rangle)_k, \tag{4.3.102}$$

where $(C\langle S \rangle)_k$ is a non-vanishing product arrangement of the $C\langle S \rangle$ term with multiplicity m_k and Σ' denotes a (possibly multiple) partial summation.

In the present application, a typical partial summation appearing in the eighth moment of $|E|$, with dispersion, is

$$\sum_i \sum_j \sum_k |n_i|^4 n_j^2 n_k^{*2} \langle |T_i|^4 \rangle \langle |T_j|^2 \rangle \langle |T_k|^2 \rangle. \tag{4.3.103}$$

In this summation, $i \neq j$, $j \neq k$ and $k \neq i$, usual designations of moments of the trigonometric structure factor are employed and n_j is the complex normalized scattering factor of the jth atom. The n-dependent coefficient $C = |n_i|^4 n_j^2 n_k^{*2}$ for this summation can be schematically written as

$$(nn^* nn^*)(nn)(n^* n^*), \tag{4.3.104}$$

the parentheses containing those ns for which the indices in the original eightfold summation were contracted. The number of different ways of placing two ns and two n^*s in the first group in (4.3.103) is

$$\binom{4}{2}^2 = 36,$$

the remaining two groups being formed in one distinct way only. Hence, the multiplicity of (4.3.103) in eqn(4.3.102) with $n = 4$ is $m = 36$. A detailed description of the partial summations appearing in the fourth, sixth and eighth moments of $|F|$ with dispersion, along with their multiplicities, has been given by Shmueli and Wilson (1983), who also completed the derivation of the non-ideal subcentric pdf of z by symbolic programming with the aid of REDUCE. It is possible that other systems, such as MACSYMA or *Mathematica*, could perform this calculation equally well or better.

As will be seen in the next chapter, a much simpler derivation of an exact representation of the subcentric pdf can be given in terms of a Fourier series. Nevertheless, the example treated above is of interest insofar as it illustrates a general application of the correction-factor method. Such a 'non-standard' correction factor can also be derived for the bicentric pdf (see section 3.3). The derivation of the moments and orthogonal polynomials for the latter case can also be carried out by computerized symbolic programming methods but the result is much more complicated and less accurate than that obtained by exact Fourier representation of the bicentric pdf as discussed in the next chapter (section 5.7).

5

FOURIER REPRESENTATIONS OF PDFS

5.1 Principle of the method

One can always express the pdf of a sum of random variables in terms of a Fourier integral containing the characteristic function as part of the integrand [as in eqn(2.1.11)]. However, an accurate numerical evaluation of such integrals can pose difficulties due to the oscillatory integrand. Consequently this approach has not been exploited in crystallographic applications. Nevertheless, a slight modification does lead to representations which are found, in practice, to produce accurate results with a reasonable amount of numerical effort.

In the approach to be discussed in some detail the Fourier integral representation of the pdf is replaced by an equivalent Fourier series. Such a replacement is possible because the structure factor, or its real or imaginary part, is necessarily a bounded quantity. Consider, for example, the normalized structure factor for the space group $P\bar{1}$. This is written

$$E(\mathbf{h}) = 2\sum_{j=1}^{N/2} n_j \cos\theta_j, \tag{5.1.1}$$

where, as before, we assume that the θs are uniformly distributed on $(-\pi, \pi)$, the atomic contributions to the structure factor are independent and dispersion and non-crystallographic symmetry are (for the time being) neglected. In the absence of dispersion we can define the maximum possible value of $E(\mathbf{h})$ to be E_{max}, where

$$E_{max} = \sum_{j=1}^{N} n_j. \tag{5.1.2}$$

Because of the form of eqn(5.1.1), values of $E^2(\mathbf{h})$ for which $E^2(\mathbf{h}) > E^2_{max}$ cannot occur. This restriction is also a constraint on the pdf, $p(E)$, which is to say that $p(E) \equiv 0$ for all values of $E^2 > E^2_{max}$. Because of this property we may expand $p(E)$ in a Fourier series with the understanding that only the values of $p(E)$ with E in the interval $[-E_{max}, E_{max}]$ are relevant.

The idea behind this mathematical trick is due to Barakat who first applied it to a model of polymer configurations (1973) and later to the theory of laser speckle (1974). This method of representing pdfs by Fourier series was applied to the problem of one- and two-dimensional random walks by Weiss and Kiefer (1983). The expansion of pdfs of one or more structure factors in Fourier series has been extensively applied to problems of crystallographic statistics by Shmueli and his collaborators, following an initial paper by Shmueli, Weiss, Kiefer and Wilson (1984) on this subject. In this chapter we discuss a number of applications of the general technique to derive Fourier-series representations of pdfs for intensity statistics. In the next chapter representations of joint pdfs of several structure factors will be derived in terms of Fourier series. These series will then be applied to a number of problems in the theory of direct methods of phase determination.

Before attempting a systematic presentation of the topic, we illustrate the basic ideas by deriving a representation of the pdf of E as a Fourier series for space group $P\bar{1}$. The one-dimensional Fourier series for any centrosymmetric space group will be written as

$$p(E) = \frac{1}{2E_{max}} \sum_{m=-\infty}^{\infty} C_m \exp\left(-\frac{\pi i m E}{E_{max}}\right). \tag{5.1.3}$$

The Fourier coefficients of this series are given by

$$C_m = \int_{-E_{max}}^{E_{max}} p(E) \exp\left(\frac{\pi i m E}{E_{max}}\right) dE. \tag{5.1.4}$$

Since $p(E)$ necessarily vanishes outside the interval $[-E_{max}, E_{max}]$, the integration limits in eqn(5.1.4) may be replaced by $\pm\infty$, with no loss of generality. We can therefore write the expression for the Fourier coefficient as

$$C_m = \int_{-\infty}^{\infty} p(E) \exp\left(\frac{\pi i m E}{E_{max}}\right) dE. \tag{5.1.5}$$

If we now compare eqn(5.1.5) with eqn(1.4.28) we see that the Fourier coefficient C_m is just the value of the characteristic function at the point $\omega = \pi m/E_{max}$, which we denote by ω_m. The The coefficient C_m is given for space group $P\bar{1}$ by

$$C_m = \langle \exp(i\omega_m E) \rangle \tag{5.1.6}$$

$$= \prod_{j=1}^{N/2} J_0(2\omega_m n_j), \tag{5.1.7}$$

where $J_0(x)$ is a zeroth-order Bessel function and the intermediate steps leading to eqn(5.1.7) are identical to those leading from eqn(2.2.23) to eqn(2.2.26). Since $J_0(0) = 1$ and $J_0(-x) = J_0(x)$, we have $C_0 = 1$ and $C_{-m} = C_m$. Equation (5.1.3) can therefore be rewritten as the Fourier series

$$p(E) = \frac{1}{2E_{max}} \left[1 + 2 \sum_{m=1}^{\infty} C_m \cos(\omega_m E) \right] . \tag{5.1.8}$$

Since $p(E)$ is a symmetric function of E, we finally have for the pdf of the magnitude of E, for space group $P\bar{1}$, the following explicit expression

$$p(|E|) = \frac{1}{E_{max}} \left\{ 1 + 2 \sum_{m=1}^{\infty} \left[\prod_{j=1}^{N/2} J_0(2\omega_m n_j) \right] \cos(\omega_m E) \right\} . \tag{5.1.9}$$

An extensive series of computations with this formula indicates that the series converges very well even for extremely heterogeneous atomic compositions. Figure 5.1 summarizes results obtained for a hypothetical structure, containing 14 carbons and one uranium in the asymmetric unit of space group $P\bar{1}$, for values of $|E|$ between 0 and 3. The histogram in Fig. 5.1 was constructed from 100 000 $|E|$ values which were obtained by direct computation from eqn(5.1.1). In this computation, θ_j for each atom and each $|E|$ value was replaced by a computer-generated pseudorandom number multiplied by 2π, thus giving an empirically derived histogram as an approximation to the pdf of $|E|$. The solid curve is a representation for the Fourier pdf of $|E|$ computed from eqn(5.1.9). The dashed curve in Fig. 5.1 is the ideal centric pdf (Wilson 1949; see also section 3.2) and the dash-dotted curve in the figure is a Gram–Charlier approximation with a four-term correction factor multiplying the ideal centric pdf of $|E|$ (Shmueli and Wilson 1981; see also section 4.3). The agreement of the Fourier pdf with the histogram is clearly excellent while the correction-factor pdf fails to cope with this rather extreme heterogeneity. The ideal pdf is given here only for comparison; this is a situation in which the central limit theorem, because of the rather small number of atoms in the unit cell, cannot be expected to give meaningful results. Note that Fig. 5.1 answers the question which we posed at the beginning of Chapter 4: since the Gaussian approximation may — under some circumstances — be qualitatively inaccurate, it is desirable to find corrections to this approximation or construct exact representations of the pdfs that can be easily evaluated. The figure illustrates both remedies to the problem, and the exact representations are obviously superior to the correction-factor approach.

Other advantages of the Fourier method are its relative simplicity [compare eqn(5.1.9) with eqn(4.3.84)] and ease of coding the expressions in a computer program. This preliminary example indicates that the Fourier method for constructing pdfs of $|E|$ is a promising one. The remainder of this chapter will be devoted to a discussion of details of the method as applied to intensity statistics.

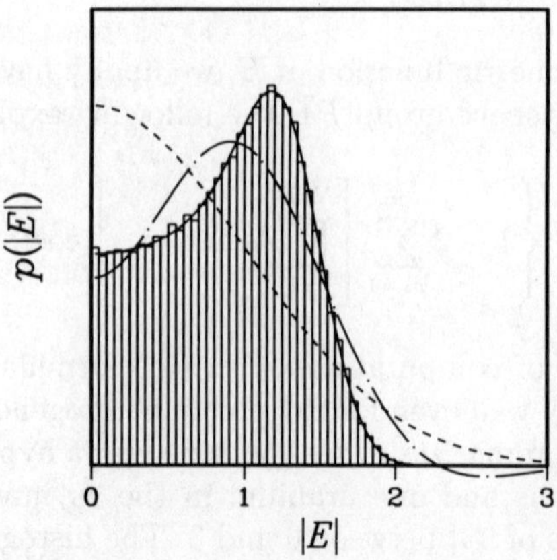

FIG. 5.1. *Comparison of theoretical and simulated pdfs of* $|E|$ *in space group* $P\bar{1}$.

5.2 General expressions of Fourier pdfs for intensity statistics

We summarize here the Fourier representations of the pdfs of $|E|$ for centrosymmetric and non-centrosymmetric space groups, restricting concrete examples to dispersionless structures with crystallographic symmetry only, and with all the atoms occupying general Wyckoff positions. These restrictions will be relaxed in later sections. As in other parts of this monograph we shall assume that all the atomic contributions to the structure factor are statistically independent, and that the atomic phase factor, $\theta_j = 2\pi(hx_j + ky_j + lz_j)$, or its components, $2\pi hx_j$, etc., are uniformly distributed over $(-\pi, \pi)$. The normalized structure factor is given by

$$E(\mathbf{h}) = \sum_{j=1}^{N/g} n_j T_j(\mathbf{h}) \equiv A(\mathbf{h}) + iB(\mathbf{h}), \qquad (5.2.10)$$

where

$$\begin{aligned} T_j(\mathbf{h}) &= \xi_j(\mathbf{h}) + i\eta_j(\mathbf{h}) \\ &= \sum_{s=1}^{g} \exp[2\pi i \mathbf{h}^T(\mathbf{P}_s \mathbf{r}_j + \mathbf{t}_s)] \end{aligned} \qquad (5.2.11)$$

is the trigonometric structure factor. In eqns (5.2.10) and (5.2.11) N is the number of atoms in the unit cell, g is the number of asymmetric units in the unit cell (i.e. order of the point group times the multiplicity of the Bravais lattice), n_j is the normalized scattering factor of the jth atom and $(\mathbf{P}_s, \mathbf{t}_s)$

is the sth space-group operator that transforms the atomic position vector, $\mathbf{r}_j$, to that of an equivalent atom.

As in the example presented in the last section, the principle behind our derivation of exact pdfs of the structure factor in the form of a Fourier series rests on two premises:

1. A centrosymmetric normalized structure factor, or the real or imaginary part of a non-centrosymmetric one, is confined to the interval $[-E_{max}, E_{max}]$, where E_{max} is the sum of the normalized scattering factors extending over the unit cell.
2. Since the probability of finding (a centrosymmetric) E or either of A and B outside $[-E_{max}, E_{max}]$ is zero, the pdfs of these quantities can be expanded in a Fourier series on this interval.

5.2.1 *Centrosymmetric space groups*

The Fourier representation of the pdf of E is given for these space groups by

$$p(|E|) = \frac{1}{E_{max}} \left[1 + 2 \sum_{m=1}^{\infty} C_m \cos(\omega_m E) \right] , \tag{5.2.12}$$

where

$$C_m = \prod_{j=1}^{N/g} C_{mj} \tag{5.2.13}$$

and

$$C_{mj} = \langle \exp(i\omega_m n_j \xi_j) \rangle \tag{5.2.14}$$

is the contribution of the jth atom to the Fourier coefficient (i.e. characteristic function), in which ξ_j is the (real) trigonometric structure factor of the jth atom. Equation (5.2.12) should be compared with eqn(5.1.9) of the previous section. Equation (5.2.12) was first applied to general space-group symmetry by Shmueli and Weiss (1987).

5.2.2 *Non-centrosymmetric space groups*

As in Chapter 3, we begin our calculation from an exact representation of the joint pdf of A and B. Since A and B are defined in the same bounded interval, $[-E_{max}, E_{max}]$, their joint pdf can be represented in terms of a two-dimensional Fourier series

$$p(A, B) = \frac{1}{4E_{max}^2} \sum_u \sum_v C_{uv} \exp\left[-\frac{\pi i(uA + vB)}{E_{max}} \right] , \tag{5.2.15}$$

where

$$C_{uv} = \prod_{j=1}^{N/g} C_{uvj} \text{ and } C_{uvj} = \left\langle \exp\left[\frac{\pi i n_j (u\xi_j + v\eta_j)}{E_{max}}\right]\right\rangle, \tag{5.2.16}$$

the trigonometric structure factors ξ_j and η_j being defined in eqn(5.2.11). The present calculation is based on the assumption that A and B are bounded by the same quantities $\pm E_{max}$. However, the case of unequally bounded A and B occurs in practice and will be considered in section 5.8.

Since, in the context of intensity statistics, we are interested in the pdf of the magnitude $|E|$ of the normalized structure factor rather than the joint pdf of A and B, we transform eqn(5.2.15) to polar coordinates by introducing the variables $|E|$ and φ, where φ is the phase of the structure factor. We have

$$p(|E|\cos\varphi, |E|\sin\varphi) = \frac{1}{4E_{max}^2}\sum_u\sum_v C_{uv}\exp\left[-\frac{\pi i|E|(u\cos\varphi + v\sin\varphi)}{E_{max}}\right] \tag{5.2.17}$$

$$= \frac{1}{4E_{max}^2}\sum_u\sum_v C_{uv}\exp\left[-\frac{iT_{uv}\cos(\varphi - \Delta)}{E_{max}}\right], \tag{5.2.18}$$

where $T_{uv} = \pi|E|(u^2 + v^2)^{1/2}$ and $\Delta = \tan^{-1}(v/u)$. If we now change the variables to $(|E|, \varphi)$ and integrate with respect to φ in each term of eqn(5.2.18) we obtain

$$p(|E|) = \frac{\pi|E|}{2E_{max}^2}\sum_u\sum_v C_{uv}J_0\left[\frac{\pi|E|(u^2 + v^2)^{1/2}}{E_{max}}\right]. \tag{5.2.19}$$

Equations (5.2.12) and (5.2.19) are applicable to all crystallographic space groups, once the Fourier coefficients are expressed in terms of the trigonometric structure factors and their averages. Equation (5.2.19) was first given by Weiss, Shmueli, Kiefer and Wilson (1985).

A considerable reduction in the computing effort required for the evaluation of the double Fourier series for the pdf (5.2.19) is possible when the Fourier coefficient C_{uv} is radially symmetric in (u, v), that is to say, it depends only on $(u^2 + v^2)^{1/2}$, rather than on u and v separately. The pdf of $|E|$ can then be represented by a *single* Fourier–Bessel series (Berg and McGregor 1966), which takes the form

$$p(|E|) = \frac{2|E|}{E_{max}^2}\sum_{k=1}^{\infty} D_k J_0\left(\frac{\gamma_k|E|}{E_{max}}\right). \tag{5.2.20}$$

In this representation

$$D_k = [J_1^2(\gamma_k)]^{-1} C\left(\frac{\gamma_k}{E_{max}}\right), \tag{5.2.21}$$

in which

$$C(\alpha\gamma_k) = \prod_{j=1}^{N/g} C_{kj}(\alpha\gamma_k), \tag{5.2.22}$$

where $J_1(x)$ is a first-order Bessel function and γ_k is the kth root of the equation $J_0(x) = 0$ (Abramowitz and Stegun 1972). A very convenient approximation for computing these roots of higher orders is given by McMahon's formula

$$\gamma_n = \beta + \frac{1}{8\beta} - \frac{124}{3(8\beta)^3} + \frac{120928}{15(8\beta)^5} - \frac{401743168}{105(8\beta)^7} + \cdots,$$

where $\beta = (n - 1/4)\pi$ (e.g. Abramowitz and Stegun, 1972), while roots of lower orders are accurately tabulated (see latter reference). Shmueli *et al.* (1984) find that the first five terms of McMahon's formula give the roots from γ_6 on with a relative error that is less than 10^{-11}.

5.3 Calculation of Fourier coefficients for different symmetries

We saw in the previous section that the construction of a pdf of $|E|$ for a given space group requires the evaluation of an average given by eqn(5.2.14) or eqn(5.2.16) for centrosymmetric or non-centrosymmetric space groups respectively. The averaging step is performed directly over the arguments of the functions appearing in the trigonometric structure factor, ξ_j or $\xi_j + i\eta_j$. This situation is analogous to the calculation of the absolute even moments of the trigonometric structure factor (section 4.2) which were also obtained by direct averaging of these arguments without requiring an expression for the exact pdf.

The atomic contributions to the Fourier and Fourier–Bessel coefficients of the pdf of $|E|$ were derived by Shmueli and Weiss (1987) for space groups of symmetries lower than tetragonal, with the exception of $Fdd2$ and $Fddd$. The work on such derivations for higher symmetries was extended by Rabinovich, Shmueli, Stein, Shashua and Weiss (1991*a*) to include all 17 plane groups and the first 206 space groups (up to and including $Fd\bar{3}$). In addition, characteristic functions were derived for all the parity groups of the reflection indices h, k and l, for which the averages have different functional forms within the same space group. We shall not reproduce all the derivations instead discussing several selected examples which serve to illustrate the methodology that has been used in the derivations.

5.3.1 *Space group P*1

The trigonometric structure factor is $\xi_j + i\eta_j$, where $\xi_j = \cos\theta_j$ and $\eta_j = \sin\theta_j$. If we substitute these values in eqn(5.2.16) we obtain

$$C_{uvj} = \left\langle \exp\left[\frac{\pi i n_j (u\cos\theta_j + v\sin\theta_j)}{E_{max}}\right]\right\rangle \tag{5.3.23}$$

$$= \left\langle \exp\left[\frac{\pi i n_j (u^2+v^2)^{1/2}\cos(\theta_j - \Delta)}{E_{max}}\right]\right\rangle \tag{5.3.24}$$

$$= \frac{1}{2\pi}\int_{-\pi}^{\pi} \exp\left[\frac{\pi i n_j (u^2+v^2)^{1/2}\cos(\theta - \Delta)}{E_{max}}\right] d\theta \tag{5.3.25}$$

$$= J_0\left[\frac{\pi n_j (u^2+v^2)^{1/2}}{E_{max}}\right], \tag{5.3.26}$$

where $\Delta = \tan^{-1}(v/u)$. The introduction of polar coordinates, leading from eqn(5.3.23) to eqn(5.3.24), and the integral representation of the Bessel function $J_0(x)$ (e.g. Abramowitz and Stegun 1972, entry 9.1.21), leading from eqn(5.3.25) to eqn(5.3.26), are frequently employed in these calculations. Equation (5.3.26) can be used in computing the Fourier coefficient for $P1$ from the two-dimensional pdf whose form is given in eqn(5.2.19). Since, however, C_{uvj} depends on $(u^2+v^2)^{1/2}$ alone, the double Fourier series can be replaced by the single Fourier–Bessel expansion in eqn(5.2.20). In this representation the coefficient C_{kj} in eqn(5.2.22) is given by $J_0(\gamma_k n_j/E_{max})$.

5.3.2 *Space group Pmmm and related symmetries*

The trigonometric structure factor has the form $\xi_j = 8\cos\theta_j\cos\varphi_j\cos\psi_j$, where $\theta_j = 2\pi h x_j$, $\varphi_j = 2\pi k y_j$ and $\psi_j = 2\pi l z_j$. These angular variables are assumed to be independent and uniformly distributed on $(-\pi,\pi)$. The atomic contribution to the Fourier coefficient is therefore given by the following integral, which we evaluate in the following sequence of equations:

$$C_{uj} = \left(\frac{1}{2\pi}\right)^3 \int_{-\pi}^{\pi}\int_{-\pi}^{\pi}\int_{-\pi}^{\pi} \exp\left(\frac{8\pi i n_j u\cos\theta\cos\varphi\cos\psi}{E_{max}}\right) d\theta d\varphi d\psi \tag{5.3.27}$$

$$= \left(\frac{1}{2\pi}\right)^2 \int_{-\pi}^{\pi}\int_{-\pi}^{\pi} J_0\left(\frac{8\pi n_j u\cos\theta\cos\varphi}{E_{max}}\right) d\theta d\varphi, \tag{5.3.28}$$

where use has been made of the integral representation of $J_0(x)$. We can further integrate on one of the remaining angular variables, by making use of the known definite integral

$$\frac{2}{\pi}\int_0^{\pi/2} J_0(2z\cos x)dx = J_0^2(z) \tag{5.3.29}$$

(Gradshteyn and Ryzhik 1980; entry 6.681-5). In this way we obtain

$$C_{uj} = \frac{1}{2\pi}\int_{-\pi}^{\pi} J_0^2\left(\frac{4\pi n_j u \cos\theta}{E_{max}}\right) d\theta = \frac{1}{2\pi}\int_{-\pi}^{\pi} J_0^2\left(\frac{4\pi n_j u \sin\theta}{E_{max}}\right) d\theta. \tag{5.3.30}$$

Further integration in closed form does not appear to be possible. We can, however, convert eqn(5.3.30) into a rapidly convergent series whose computation is much more efficient than the numerical integration that might otherwise be necessary. For this purpose we substitute the relation

$$J_0(z\sin\beta) = \sum_{k=-\infty}^{\infty} J_k^2(z/2)\exp(ik\beta) \tag{5.3.31}$$

(Gradshteyn and Ryzhik 1980; entry 8.531-3) into eqn(5.3.30), interchange the order of the summation and the integration and arrive at

$$C_{uj} = \sum_{k=-\infty}^{\infty}\sum_{l=-\infty}^{\infty} J_k^2\left(\frac{2\pi n_j u}{E_{max}}\right) J_l^2\left(\frac{2\pi n_j u}{E_{max}}\right)\left\{\frac{1}{2\pi}\int_{-\pi}^{\pi}\exp[i(k+l)\theta]d\theta\right\}. \tag{5.3.32}$$

However, the integral over θ in eqn(5.3.32) is the Kronecker delta, $\delta_{k,-l}$, which allows us to write

$$C_{uj} = \sum_{k=-\infty}^{\infty}\sum_{l=-\infty}^{\infty} J_k^2\left(\frac{2\pi n_j u}{E_{max}}\right) J_l^2\left(\frac{2\pi n_j u}{E_{max}}\right)\delta_{k,-l} \tag{5.3.33}$$

$$= \sum_{k=-\infty}^{\infty} J_k^4\left(\frac{2\pi n_j u}{E_{max}}\right) \tag{5.3.34}$$

$$= J_0^4\left(\frac{2\pi n_j u}{E_{max}}\right) + 2\sum_{k=1}^{\infty} J_k^4\left(\frac{2\pi n_j u}{E_{max}}\right). \tag{5.3.35}$$

The final result, eqn(5.3.35), converges very quickly.

The considerations discussed here are important in similar calculations for all the orthorhombic, some tetragonal and all the cubic space groups isomorphic to the point groups 23 and $m\bar{3}$.

5.3.3 *Plane group p*3

The derivation of the pdf for this plane group illustrates the methodology used in derivations that involve space groups or plane groups containing a unique threefold or a sixfold axis. The normalized structure factor for the plane group $p3$ is given by $E = A + iB$, where

$$A = \sum_{j=1}^{N/3} n_j \sum_{k=1}^{3}\cos\mu_{jk} \text{ and } B = \sum_{j=1}^{N/3} n_j \sum_{k=1}^{3}\sin\mu_{jk} \tag{5.3.36}$$

with

$$\begin{aligned} \mu_{j1} &= 2\pi(hx_j + ky_j) \\ \mu_{j2} &= 2\pi(kx_j + iy_j) \\ \mu_{j3} &= 2\pi(ix_j + hy_j), \end{aligned} \tag{5.3.37}$$

where $i = -h-k$ (*International Tables for X-ray Crystallography* 1965), not to be confused with $i = \sqrt{-1}$. Note that only two of the μs are independent, the remaining one being found from the relation $\mu_{j1} + \mu_{j2} + \mu_{j3} = 0$.

The characteristic function for the pdf of E, required for the coefficients of the two-dimensional Fourier series, is given by

$$\begin{aligned} C(\omega_1, \omega_2) &= \langle \exp[i(\omega_1 A + \omega_2 B)] \rangle \\ &= \prod_{j=1}^{N/3} \left\langle \exp\left[in_j \sum_{k=1}^{3} (\omega_1 \cos \mu_{jk} + \omega_2 \sin \mu_{jk}) \right] \right\rangle, \end{aligned} \tag{5.3.38}$$

where use has been made of the assumption of independence of the atomic contributions to the structure factor. If we set $\omega_1 = \omega \sin \Delta$ and $\omega_2 = \omega \cos \Delta$, so that $\Delta = \tan^{-1}(\omega_1/\omega_2)$ and $\omega = (\omega_1^2 + \omega_2^2)^{1/2}$, we obtain

$$C(\omega, \Delta) = \prod_{j=1}^{N/3} \left\langle \exp\left[in_j\omega \sum_{k=1}^{3} (\sin \Delta \cos \mu_{jk} + \cos \Delta \sin \mu_{jk}) \right] \right\rangle \tag{5.3.39}$$

$$= \prod_{j=1}^{N/3} \left\langle \exp\left[in_j\omega \sum_{k=1}^{3} \sin(\mu_{jk} + \Delta) \right] \right\rangle \tag{5.3.40}$$

$$\equiv \prod_{j=1}^{N/3} C_j(\omega, \Delta). \tag{5.3.41}$$

Since the atomic phase factors are assumed to be uniformly distributed we may rewrite the average in eqn(5.3.40) as

$$\begin{aligned} C_j(\omega, \Delta) = \frac{1}{(2\pi)^2} \int_{-\pi}^{\pi} \int_{-\pi}^{\pi} & \exp\{in_j\omega[\sin(\mu_1 + \Delta) + \sin(\mu_2 + \Delta) \\ & + \sin(-\mu_1 - \mu_2 + \Delta)]\} d\mu_1 d\mu_2. \end{aligned} \tag{5.3.42}$$

The integral in eqn(5.3.42) can be computed numerically, but it is possible and much more efficient to transform it into a rapidly converging series. This can be done using the identity

$$\exp(ix \sin \beta) = \sum_{k=-\infty}^{\infty} J_k(x) \exp(ik\beta) \tag{5.3.43}$$

(e.g. Gradshteyn and Ryzhik 1980), where $J_k(x)$ is a Bessel function of order k. If we make use of eqn(5.3.43) we can rewrite eqn(5.3.42) as

$$C_j(\omega, \Delta) = \sum_{k=-\infty}^{\infty} \sum_{m=-\infty}^{\infty} \sum_{n=-\infty}^{\infty} J_k(n_j\omega)J_m(n_j\omega)J_n(n_j\omega) \exp[i(k+m+n)\Delta]$$

$$\times \frac{1}{(2\pi)^2} \int_{-\pi}^{\pi} \int_{-\pi}^{\pi} \exp\{i[(k-n)\mu_1 + (m-n)\mu_2]d\mu_1 d\mu_2 \qquad (5.3.44)$$

$$= \sum_{k=-\infty}^{\infty} \exp(3ik\Delta) J_k^3(n_j\omega) \qquad (5.3.45)$$

since the value of the integral in eqn(5.3.44) is the product of Kronecker deltas, $\delta_{kn}\delta_{mn}$. The corresponding Fourier coefficient is formed by replacing ω_1 with $\pi\alpha u$ and ω_2 with $\pi\alpha v$, or ω with $\pi\alpha(u^2+v^2)^{1/2}$ and Δ with $\tan^{-1}(u/v)$. In view of the presence of Δ in the expression for the characteristic function, the pdf required in this case must be that based on the double Fourier series, i.e. eqn(5.2.19).

The technique applied in this example will also be applied in the next chapter to the analysis of some invariants and semi-invariants used in direct methods.

5.3.4 *Space group P6/mcc*

In this example we deal with a three-dimensional space group of rather high symmetry and illustrate the use of the periodic Dirac delta function (e.g. Lighthill 1964) in the treatment of dependent random variables. We derive the characteristic function for the hexagonal space group $P6/mcc$, and for reflection indices hkl with $l = 2n+1$. If the space group origin is chosen at a centre of symmetry, the real part of the normalized structure factor is given by

$$A = 4\sum_{j=1}^{24} n_j \cos\varphi_j \sum_{k=1}^{3} (\cos\alpha_{jk} - \cos\beta_{jk}) \qquad (5.3.46)$$

and the imaginary part vanishes identically. In this equation,

$$\alpha_{j1} = 2\pi(hx_j + ky_j),\ \alpha_{j2} = 2\pi(kx_j + iy_j),\ \alpha_{j3} = 2\pi(ix_j + hy_j),$$

$$\beta_{j1} = 2\pi(kx_j + hy_j),\ \beta_{j2} = 2\pi(hx_j + iy_j),\ \beta_{j3} = 2\pi(ix_j + ky_j)$$

and $\varphi_j = 2\pi l z_j$. Note that $\sum_{k=1}^{3} \alpha_{jk} = 0$ and $\sum_{k=1}^{3} \beta_{jk} = 0$, that is, the problem of dependence already encountered in the derivation of the characteristic function for the plane group $p3$ is also present here. The trigonometric structure factor used in eqn(5.3.46) is taken from the recent version of the structure factor tables (Shmueli 1993).

The required characteristic function, expressed as a product of atomic contributions, is

$$C(\omega_1) = \prod_{j=1}^{24} \left\langle \exp\left[4in_j\omega_1 \cos\varphi_j \sum_{k=1}^{3}(\cos\alpha_{jk} - \cos\beta_{jk})\right]\right\rangle. \qquad (5.3.47)$$

The atomic characteristic function $C_j(\omega_1)$ is obtained by evaluating the average in eqn(5.3.47) over all the angular variables involved. We can do it, as in the previous subsection, by solving for α_3 explicitly as $(-\alpha_1 - \alpha_2)$, and with a similar expression for β. However, one can also give a derivation which retains the inherent symmetry by introducing periodic Dirac delta functions of the sums of the αs and βs. The periodic Dirac delta function can be represented as

$$\delta_{2\pi}(x) \equiv \sum_{m=-\infty}^{\infty} \delta(x - 2\pi m) = \frac{1}{2\pi}\sum_{k=-\infty}^{\infty} \exp(-ikx) \qquad (5.3.48)$$

(e.g. Lighthill 1964; p.67). It follows that $\delta_{2\pi}(\alpha_1 + \alpha_2 + \alpha_3)$ vanishes everywhere except when the sum of the three αs is equal to zero. When we insert the appropriate version of eqn(5.3.48) into the expression for $C_j(\omega_1)$ we find that this function is given by

$$\begin{aligned} C_j(\omega_1) = {} & \frac{1}{(2\pi)^5}\int_{-\pi}^{\pi} d\varphi \int_{-\pi}^{\pi}\int_{-\pi}^{\pi}\int_{-\pi}^{\pi}\int_{-\pi}^{\pi}\int_{-\pi}^{\pi}\int_{-\pi}^{\pi} d\alpha_1 d\alpha_2 d\alpha_3 d\beta_1 d\beta_2 d\beta_3 \\ & \times \delta_{2\pi}(\alpha_1 + \alpha_2 + \alpha_3)\delta_{2\pi}(\beta_1 + \beta_2 + \beta_3) \\ & \times \exp\left[4in_j\omega_1 \cos\varphi \sum_{k=1}^{3}(\cos\alpha_k - \cos\beta_k)\right]. \qquad (5.3.49) \end{aligned}$$

The integrand can be factored into a product of two functions, one depending on the αs and independent of the βs and the other depending on the βs and independent of the αs. Hence we can express $C(\omega_1)$ as

$$C_j(\omega_1) = \frac{1}{2\pi}\int_{-\pi}^{\pi} d\varphi \mathcal{P}_j \mathcal{Q}_j, \qquad (5.3.50)$$

where $\mathcal{P}_j$ and $\mathcal{Q}_j$ are the α-dependent and β-dependent functions respectively. If we make use of the representation of the Dirac delta function given in eqn(5.3.48) we have

$$\begin{aligned} \mathcal{P}_j = {} & \frac{1}{(2\pi)^2}\int_{-\pi}^{\pi}\int_{-\pi}^{\pi}\int_{-\pi}^{\pi} d\alpha_1 d\alpha_2 d\alpha_3 \delta_{2\pi}(\alpha_1 + \alpha_2 + \alpha_3) \\ & \times \exp\left[4in_j\omega_1 \cos\varphi \sum_{k=1}^{3} \cos\alpha_k\right] \end{aligned}$$

$$= \sum_{k=-\infty}^{\infty} \left[\frac{1}{2\pi} \int_{-\pi}^{\pi} \exp(-ik\alpha + 4in_j\omega_1 \cos\varphi \cos\alpha) d\alpha \right]^3 . \quad (5.3.51)$$

The introduction of the new variable $\alpha' = \alpha + \frac{\pi}{2}$ reduces the expression in the square brackets in eqn(5.3.51) to a standard representation of the Bessel function $J_k(x)$ multiplied by a phase factor. We obtain

$$\mathcal{P}_j = \sum_{k=-\infty}^{\infty} \exp\left(3ik\frac{\pi}{2}\right) \left[\frac{1}{2\pi} \int_{-\pi}^{\pi} \exp(-ik\alpha' + 4in_j\omega_1 \cos\varphi \sin\alpha') d\alpha' \right]^3 \quad (5.3.52)$$

and finally (cf. Gradshteyn and Ryzhik 1980; entry 8.411-1)

$$\mathcal{P}_j = \sum_{k=-\infty}^{\infty} \exp\left(3ik\frac{\pi}{2}\right) J_k^3(4n_j\omega_1 \cos\varphi). \quad (5.3.53)$$

A similar calculation leads to

$$\mathcal{Q}_j = \sum_{m=-\infty}^{\infty} \exp\left(-3im\frac{\pi}{2}\right) J_m^3(4n_j\omega_1 \cos\varphi). \quad (5.3.54)$$

The characteristic function for the pdf of E, corresponding to the space group and index parity, is therefore given by eqn(5.3.47), where the atomic characteristic functions are computed from eqns (5.3.50) and (5.3.53). Equation (5.3.54) is not needed since it is just the complex conjugate of eqn(5.3.53). This also shows that the atomic characteristic function is real, as $\mathcal{P}_j\mathcal{Q}_j = |\mathcal{P}_j|^2$ in the present case.

A similar calculation leads to the characteristic function for the space group $P6/mcc$ and hkl reflection indices in which l is restricted to be even. We start from a trigonometric structure factor analogous to that in eqn(5.3.46) except for the sign of $\cos\beta_{jk}$ which is now positive rather then negative (Shmueli 1993). A rapid inspection of the derivation presented in this subsection shows that the consequence of this sign change is that $\mathcal{P}_j$ and $\mathcal{Q}_j$ are identical for $P6/mcc$ with $l = 2n$, rather than complex conjugates as in $P6/mcc$ with $l = 2n+1$. We therefore see that the pdf depends not only on the space-group symmetry of the crystal but also on the parity of the reflection indices, especially in higher symmetries. We shall see some examples of this dependence in the next section.

5.4 Comparisons of exact and approximate pdfs of $|E|$

The last section illustrated derivations of characteristic functions for Fourier pdfs with various degrees of complexity. In this section we com-

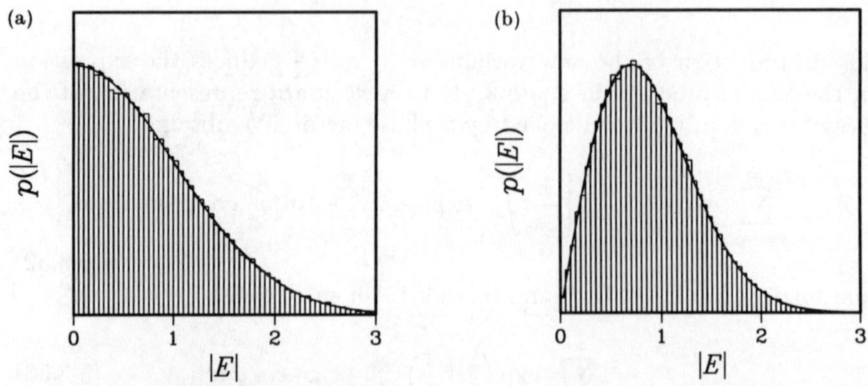

FIG. 5.2. *Comparison of theoretical pdfs of* $|E|$ *for the equal-atom case.* (a) $P\overline{1}$, $N = 40$. (b) $P6cc$, $N = 240$.

pare the performance of pdfs expressed as Fourier or Fourier–Bessel series with approximations such as those of Gram–Charlier type or those based on the central limit theorem.

We shall first illustrate the performance of the various pdfs for the equal-atom case. Figure 5.2 illustrates this for space groups $P\overline{1}$ and $P6ccc$ which are representative examples of low and high symmetries. A moderately small number of 20 equal atoms in the asymmetric unit has been assumed in generating the data.

As might have been expected, the performances of pdfs based on the central limit theorem, pdfs with a four-term correction factor and exact Fourier pdfs are remarkably similar, although the agreement of the Fourier pdfs with the histogram of simulated $|E|$ values is clearly the best of the three. The reason for this is probably the requirement of the central limit theorem that N be large. When N is large the correction factors for the Gram–Charlier or Edgeworth series will be small. It follows that, for structures composed of a reasonably large number of not too dissimilar atoms, pdfs based on the central limit theorem suffice to resolve space-group ambiguities involving an initially unknown status of centrosymmetry. These and other examples, not shown here, suggest that this is the case for all crystallographic space groups.

We shall next consider the effects of atomic heterogeneity on intensity statistics. A good qualitative indicator of the magnitude of such effects is the exact fourth absolute moment of E, discussed and displayed in section 4.2. Figures 4.2 and 4.3 show that the strongest effects are expected to appear in the triclinic space groups, although those for high tetragonal and hexagonal space groups may also be very significant. The deviations of the values of the fourth absolute moment of $|E|$ from their ideal values will serve as a guide in the choice of illustrative examples. The tests of new pdfs, to be illustrated in the exposition to follow, will be performed against distributions recalculated from published structures (see Shmueli 1982; Shmueli *et al.* 1984), or against simulated distributions. In the latter we make use of space-group specific expressions for trigonometric structure factors, such as those in *International Tables* (1965, 1993), and are able to simulate the pdf of $|E|$ for any space group and any parity of the reflection indices that give rise to a different functional form of the trigonometric structure factor.

The simulations in the following examples are based on 100 000 values of $|E|$ each, ranging between 0 and 3, and a strongly heterogeneous atomic composition. The composition of the asymmetric unit chosen corresponds to $C_{14}U$. The simulated distribution is represented by a frequency histogram of $|E|$ values, pdfs of $|E|$ based on the central limit theorem of $|E|$ (section 3.2) are denoted by dashed curves, pdfs based on the correction expansions in terms of orthogonal polynomials (section 4.3) are given by dotted-dashed curves and the Fourier or Fourier–Bessel pdfs of $|E|$ (this chapter) by solid curves. The theoretical pdfs of $|E|$ are computed at the midpoints of the histogram bins, and their values are scaled so that they are comparable with the simulated distributions corresponding to the same composition and symmetry.

Probably the most conspicuous discrepancies caused by atomic heterogeneity between pdfs based on the central limit theorem and exact ones occur in the space groups $P1$ and $P\bar{1}$ (see Fig. 5.1). As the degree of symmetry increases the discrepancies decrease, but still remain significant in many space groups of higher symmetry. This is illustrated in Figs 5.4 and 5.5, which compare various representations of the pdfs for the space groups $I4_1md$ and $P6_1$ respectively. Again, the Fourier or Fourier–Bessel pdfs agree very well with the histograms of simulated values, while the performance of the correction-factor pdfs, although better than that of the ideal pdfs (i.e. those based on the central limit theorem), is significantly poorer.

An illustrative example for triclinic space groups is given in Fig. 5.3, in which the histogram is recalculated from published positional and thermal parameters of a chloroplatinate ($C_6N_4O_4Cl_2Pt$, $P\bar{1}$, $Z = 2$, Faggiani, Lippert and Lock 1980). The process of recalculation consisted of (i) calculation of the independent magnitudes of conventional structure factors within the copper sphere, (ii) construction of a Wilson plot, and (iii) calculation of the magnitudes of normalized structure factors and construction of the histogram of $|E|$s shown in Fig. 5.3. The angular dependence induced by this process was approximately accounted for by computing the normalized scattering factors and reciprocal of the maximum value of $|E|$ as weighted averages, the weights being given by the numbers of (generated)

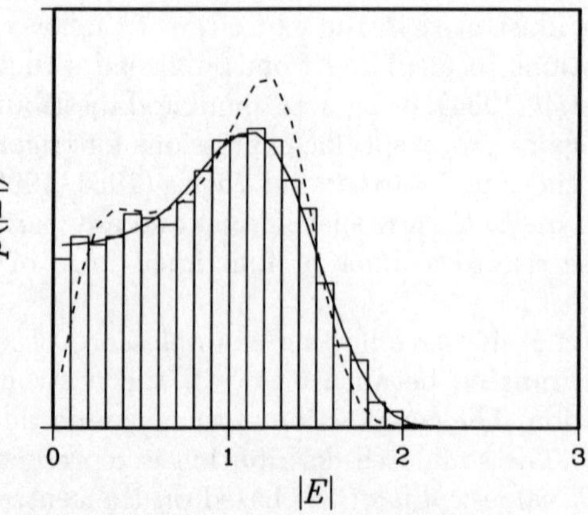

FIG. 5.3. *Theoretical pdfs of $|E|$ for $P\bar{1}$ and $P1$ and a recalculated histogram of $|E|$s for a published $P\bar{1}$ structure.* For explanations see text.

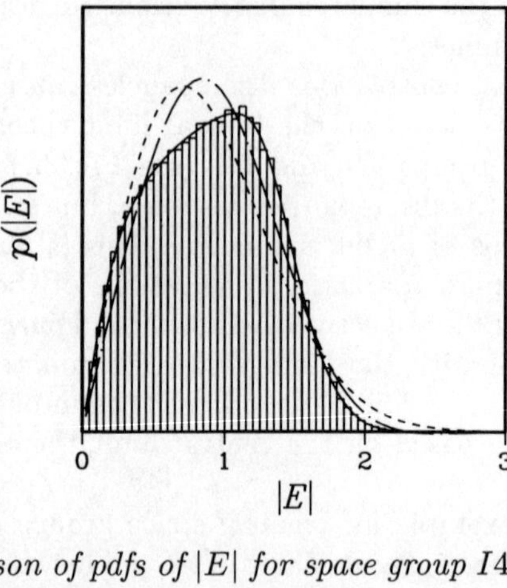

FIG. 5.4. *Comparison of pdfs of $|E|$ for space group $I4_1md$.* The reflection subset has the form *hkl*: $2k + l = 2n + 1$.

reflections in successive $\sin\theta/\lambda$ shells in the Wilson plot, constructed for this structure (Shmueli 1982; Shmueli *et al.* 1984). These average values of n_j and α were used to compute the pdf of $|E|$ for $P\bar{1}$ and $P1$, shown in Fig. 5.3 by solid and dashed curves respectively. The $P\bar{1}$ versus $P1$ space-group ambiguity therefore appears to be resolvable, even in the presence of considerable atomic heterogeneity.

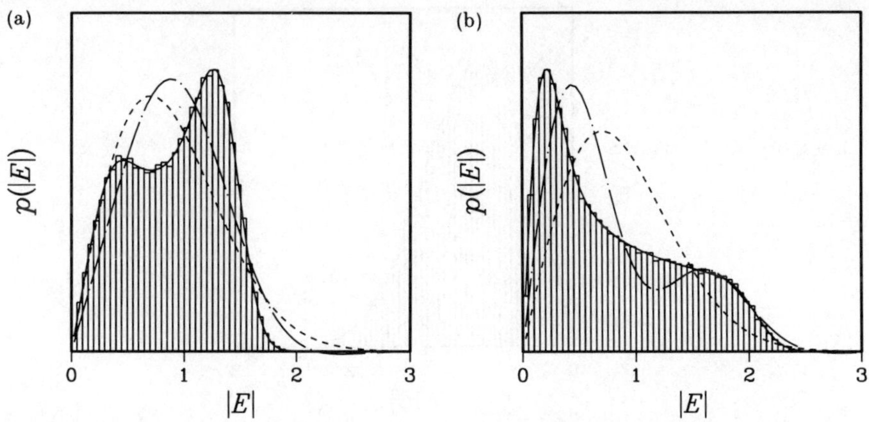

FIG. 5.5. *Comparison of pdfs of* $|E|$ *for space group* $P6_1$. (a) *hkl*: $l = 6n+1$. (b) *hkl*: $l = 6n+3$.

Figure 5.5 also illustrates another aspect of structure-factor statistics, already mentioned in subsection 5.3.4. In some space groups subsets of reflections with different parities of the *hkl* indices have different functional forms of the trigonometric structure factors and also, often, different moments of $|T|$ (Wilson 1978; Shmueli and Kaldor 1981) and different Fourier or Fourier–Bessel coefficients (Rabinovich *et al*, 1991*a*, 1991*b*). This leads to differences apparent both in the simulated and exact pdfs. For example, the space group $P6_1$ has four such subsets (*hkl*: $l = 6n, l = 6n+1(5), l = 6n+2(4)$ and $l = 6n+3$), two of which are illustrated in Fig. 5.5.

To summarize, the effects of atomic heterogeneity on intensity statistics in space groups other than cubic and under the assumption that all the atoms are located in general positions depend on the space-group symmetry, as well as on the parity of the reflection indices. This property is already apparent in the values of the absolute even moments of the trigonometric structure factor (Wilson 1978; Shmueli and Kaldor 1981,1983). Since the present calculation of the Fourier representations of the pdfs rests on the values of the characteristic functions, and these can be regarded as moment generating functions, the dependence of the Fourier representations of the pdfs on the parity of the reflection indices is consistent with the known properties of the moments.

Preliminary comparisons of the simulated distributions of $|E|$ for the cubic space groups with the corresponding pdfs based on the central limit theorem show remarkably small discrepancies, in spite of the highly heterogeneous asymmetric unit used (Rabinovich *et al.* 1991*b*). Since, when the

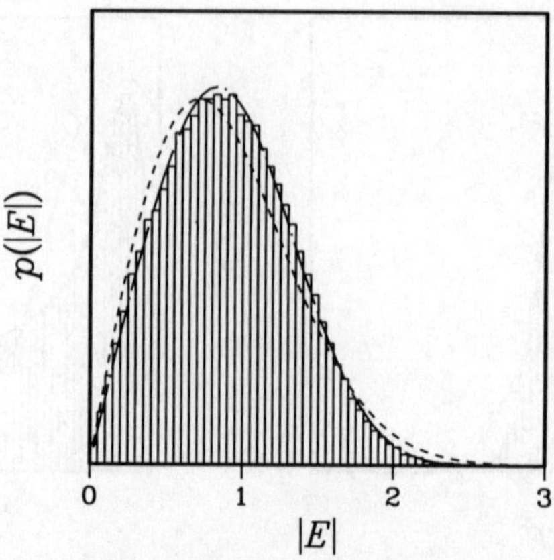

FIG. 5.6. *Comparison of pdfs of* $|E|$ *for space group* $P4_132$. Only the correction-factor (dashed-dotted), central limit theorem (dashed) and simulated (histogram) densities are shown.

exact $p(x)$ deviates only slightly from the approximate $p_0(x)$ in eqn(4.1.7), the correction-factor method is believed to be quite accurate, this method was applied to those cubic space groups for which, as pointed out above, no Fourier or Fourier–Bessel coefficients have been worked out in detail. These are the space groups isomorphous to the point groups 432, $\bar{4}3m$ and $m\bar{3}m$, for which the first four even moments of $|T|$ have been published (Shmueli and Kaldor 1981, 1983). For these the Fourier or Fourier–Bessel coefficients have rather complicated functional forms whose evaluation requires considerable computational effort. Figures 5.6 and 5.7 illustrate such comparisons of frequency histograms of $|E|$ with pdfs based on the central limit theorem and correction-factor pdfs of $|E|$ for selected index parities in space groups $P4_132$ and $Ia\bar{3}d$ respectively. It is evident that the correction-factor method can account quite satisfactorily for effects of atomic heterogeneity for these high symmetries. The situation is also very similar for all the other cubic space groups which are isomorphous to the three highest cubic point groups mentioned earlier in this paragraph.

5.5 Remarks on numerical implementations

The practical considerations required to implement formalisms based on Fourier series include some aspects of simple numerical analysis. Three types of problems occur: (i) numerical integration, (ii) computation of Bessel functions and (iii) generation of pseudorandom numbers. There is a vast number of solutions to these problems in terms of professionally written software, and we shall therefore content ourselves with some comments

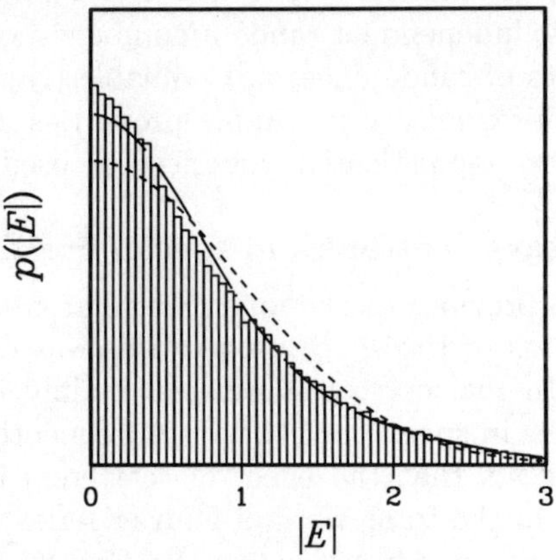

FIG. 5.7. *Comparison of pdfs of* $|E|$ *for space group* $Ia\overline{3}d$. Only the correction-factor (dashed-dotted), central limit theorem (dashed) and simulated (histogram) densities are shown.

related to our particular applications.

Numerical integration. Characteristic functions for space groups of some high symmetries must be evaluated by numerical integration (see, for example, subsection 5.3.4). The integrands often depend on Bessel functions with trigonometric arguments, and are therefore fairly oscillatory. A recommended method for the integration of such functions is the Romberg algorithm (Davis and Rabinovitz 1967), which automatically determines the extent and mode of subdivision of the range of integration into subintervals so that the integrand in each of them is as smooth as possible and can be evaluated accurately by any of the standard methods of numerical integration. We used the Romberg routine DCADRE, of the IMSL library of mathematical subroutines, on mainframe and workstation, and similar routines from the Fortran library of *Numerical Recipes* (Press, Flannery, Teukolsky and Vetterling 1990), on personal computers.

Computation of Bessel functions. A preferred way of computing arrays of Bessel functions appears to be by the use of recurrence formulae, where the program decides whether forward or backward recursion has to be employed according to the value of the argument. Most libraries of mathematical routines contain programs that operate in this way.

Random number generators. These are available in any library of programs, but not all are 'equally random'. It is therefore sometimes desirable to test their performance, even though the present simulations are not very extensive. We mention here two tests, which are particularly simple to program. One of them, the moment test, uses the fact that the nth moment of a number, intended to be uniformly distributed over [0,1], should be very close to $(n+1)^{-1}$. Another test is that of correlation. The correlation coefficient of any two independent random numbers should be small. Many other statistical tests of randomness are available (Niederreiter 1991) but our application requires only very simple properties of random numbers and therefore no more sophisticated tests need be used.

5.6 Effects of heavy scatterers in special positions

We have assumed in previous discussions that all atoms are located in general Wyckoff positions of the space group. This was done mainly for the sake of simplicity. In real crystal structures, mainly the inorganic ones, the presence of atoms in special positions is a frequently encountered phenomenon. This suggests that the effect of scatterers in special positions should be discussed in the framework of Fourier series representations.

Formal expressions for corrections to the Gaussian approximation to the pdf of the magnitude of the structure factor, allowing for presence of atoms in special positions, are given by Karle and Hauptman (1953) and by Hauptman and Karle (1953*a*). Moments of normalized intensity that incorporate dependence on atoms in special positions have been calculated by Foster and Hargreaves (1963). However, no application appears to have been made of these results to the construction of space-group dependent corrections to the Gaussian approximations to the pdf of the magnitude of the structure factor. Such an application is in principle feasible but it seems to have been discouraged by the algebraic complexity inherent in the required derivations. Furthermore, there is a large number of possible different special positions within the 230 crystallographic space groups, again militating against the enumeration of correction factors. We have examined the problem of special positions in terms of Fourier representations of the relevant pdfs (Shmueli and Weiss 1988), for most triclinic, monoclinic and orthorhombic space groups. The results are both simple and accurate as will be seen from the following discussion.

A distinction should be made between two types of special positions: (i) fixed, e.g. centres of symmetry, and (ii) variable, e.g. axes of rotation and planes of reflection. The components of the position vectors terminating at the fixed positions exhibit strong rational dependence, and none of the methods discussed so far is adequate for their analysis. From a more practical viewpoint, fixed special positions can often be inferred from the composition of the crystal and their contribution to the structure factor simply subtracted. Many examples of this can be found in the early crystallographic literature (e.g. Sim 1958). We shall therefore confine our treatment to special positions having one or two degrees of freedom only.

We start from a general representation of the normalized structure factor, which accounts for all possible special positions. We have

$$E(\mathbf{h}) = \sum_{m=1}^{k} a_m \sum_{j=1}^{N_m} n_{jm}(\xi_{jm} + i\eta_{jm}) \equiv A(\mathbf{h}) + iB(\mathbf{h}), \qquad (5.6.55)$$

where the index m ranges over the different types of Wyckoff positions, their number being k. In this equation the index j ranges over the N_m atoms in the mth Wyckoff position, a_m is the occupancy factor of the mth Wyckoff position and $\xi_{jm} + i\eta_{jm}$ is the conventional trigonometric structure factor of the jth atom in the mth Wyckoff position. The occupancy factor, a_m, is the contribution of an atom located in the mth position to the asymmetric unit. For example, the occupancy factor for an atom located on an n-fold axis is $1/n$, and $a_m = 1$ corresponds to a general position.

The characteristic function for $E(\mathbf{h})$ in the centrosymmetric case can be calculated as

$$C(\omega_1) = \langle \exp(i\omega_1 A) \rangle \qquad (5.6.56)$$

$$= \left\langle \exp\left(i\omega_1 \sum_{m=1}^{k} a_m \sum_{j=1}^{N_m} n_{jm}\xi_{jm} \right) \right\rangle \qquad (5.6.57)$$

$$= \prod_{m=1}^{k} \prod_{j=1}^{N_m} \langle \exp(i\omega_1 a_m n_{jm} \xi_{jm}) \rangle. \qquad (5.6.58)$$

Hence the evaluation of $C(\omega_1)$ reduces to calculation of the indicated average.

Let us illustrate this by an example. Consider the space group $P2/m$ in the second setting and assume that its asymmetric unit has 14 carbons and a single uranium in general positions, and one additional uranium at a general y coordinate on the twofold axis passing through the origin. Of course, only half of this uranium atom belongs to the asymmetric unit. The general form of the trigonometric structure factor is

$$\xi_j = 4\cos[2\pi(hx_j + lz_j)]\cos(2\pi k y_j) \qquad (5.6.59)$$

(*International Tables* 1965) and the corresponding average for atoms in general positions ($a_m = 1$) to be used in eqn(5.6.58) is

$$C_{j,gen} = \frac{1}{(2\pi)^2} \int_{-\pi}^{\pi} \int_{-\pi}^{\pi} \exp(4i\omega_1 n_j \cos\theta \cos\varphi) d\theta d\varphi \qquad (5.6.60)$$

$$= \frac{1}{2\pi} \int_{-\pi}^{\pi} J_0(4\omega_1 n_j \cos\theta) d\theta \qquad (5.6.61)$$

$$= J_0^2(2\omega_1 n_j). \qquad (5.6.62)$$

The average in eqn(5.6.58) for the atom located on the twofold axis ($a_m = \frac{1}{2}$), at $(0, y_j, 0)$, is given by

$$C_{j,spec} = \frac{1}{2\pi}\int_{-\pi}^{\pi} \exp(i\omega_1 n_j \frac{1}{2} 4\cos\varphi)d\varphi \tag{5.6.63}$$
$$= J_0(2\omega_1 n_j). \tag{5.6.64}$$

Notice that by directly substituting the coordinates of a special position into the general expression for the trigonometric structure factor and using the occupancy factor we avoid the need for separate expressions of this factor for each special position.

The non-centrosymmetric case is treated similarly. The characteristic function for the joint pdf of A and B is given by

$$C(\omega_1, \omega_2) = \langle \exp[i(\omega_1 A + \omega_2 B)] \rangle \tag{5.6.65}$$
$$= \left\langle \prod_{m=1}^{k} \prod_{j=1}^{N_m} \exp[i a_m n_{jm}(\omega_1 \xi_{jm} + \omega_2 \eta_{jm})] \right\rangle \tag{5.6.66}$$
$$= \prod_{m=1}^{k} \prod_{j=1}^{N_m} C_{jm}, \tag{5.6.67}$$

where

$$C_{jm} = \langle \exp[i a_m n_{jm}(\omega_1 \xi_{jm} + \omega_2 \eta_{jm})] \rangle. \tag{5.6.68}$$

Consider as an example the space group $Pmm2$. The general expressions for the components of the trigonometric structure factor are

$$\xi_j = 4\cos 2\pi h x_j \cos 2\pi k y_j \cos 2\pi l z_j \tag{5.6.69}$$
$$\eta_j = 4\cos 2\pi h x_j \cos 2\pi k y_j \sin 2\pi l z_j \tag{5.6.70}$$

(*International Tables* 1965). For an atom located on, say, the mirror plane passing through the y and z axes, the components become

$$\xi_j = 4\cos 2\pi k y_j \cos 2\pi l z_j \tag{5.6.71}$$
$$\eta_j = 4\cos 2\pi k y_j \sin 2\pi l z_j \tag{5.6.72}$$

and $a_m = \frac{1}{2}$ has to be used with eqn(5.6.68). Finally, if an atom is located on the twofold axis passing through the origin, the components of the trigonometric structure factor become

$$\xi_j = 4\cos 2\pi l z_j \tag{5.6.73}$$
$$\eta_j = 4\sin 2\pi l z_j \tag{5.6.74}$$

and the corresponding a_m is equal to $\frac{1}{4}$, since the atom is both rotated and reflected into itself (the twofold axis is a line of intersection of two perpendicular mirror planes). Carrying out calculations analogous to those in subsections 5.3.1 and 5.3.2 we obtain the following expressions for the atomic characteristic functions for the space group $Pmm2$:

Atomic characteristic function	Description of position
$\sum_{k=-\infty}^{\infty} J_k^4(\omega n_j)$	General
$J_0^2(\omega n_j)$	Mirror plane
$J_0(\omega n_j)$	Twofold axis

Atomic characteristic functions for the special positions in the space groups $P2$, $P222$ and $Pmmm$ have been derived by Shmueli and Weiss (1988). However, the pdf of $|E|$ computed from the result for $P222$ was incorrect since the trigonometric structure factor for general positions is complex while that for the twofold axes is real. This results in unequal bounds on the components A and B therefore requiring a different form of the Fourier pdf. We treat this problem in some generality in section 5.8, along with the cases of dispersion and the subcentric pdf, which are analogous to the present one in that the bounds of A and B differ.

The extent to which heavy scatterers in special positions can affect the pdf of $|E|$ may be rather dramatic. This has been illustrated by Shmueli and Weiss (1988), but can also be inferred from the examples just discussed. Suppose that a uranium is located on a twofold axis of the space group $Pmm2$, and there are 15 light atoms in the asymmetric unit of this space group, all located in general positions. The atomic characteristic function for the (uranium) on the twofold axis in $Pmm2$ is the same as that for an atom in a general position of $P1$ (see subsection 5.3.1). Since the uranium on the twofold axis is expected to dominate the scattering from the crystal, its atomic characteristic function will be the dominant contributor to the Fourier coefficient. Therefore, the pdf for such a crystal, which belongs to the space group $Pmm2$, will take on the form of the pdf for a crystal belonging to the space group $P1$, with all the atoms in general positions. Similarly, if the heavy atom is located on a mirror plane of $Pmm2$, the corresponding pdf will resemble that for the space group $P2$, with all the atoms in general positions (Shmueli and Weiss 1988).

5.7 Fourier pdfs allowing for non-crystallographic symmetry

Representations of pdfs of the structure factor in terms of Fourier series can also be applied to more general problems than those treated in earlier sections of this chapter. An extensively studied example is the not infrequent presence of non-crystallographic symmetry in the asymmetric unit. The most important class of such symmetry appears to be non-crystallographic centrosymmetry. We have rederived in section 3.3 the ideal bicentric pdf, first given in approximate form by Lipson and Woolfson (1952) and Rogers and Wilson (1953). At the end of section 4.4 we considered the feasibility of constructing correction terms for this approximation to the pdf. We realized quite soon after the first application of the Fourier method (Shmueli *et al.* 1984) that non-crystallographic symmetry can be rather easily treated with this new tool (Shmueli, Weiss and Kiefer 1985), and therefore summarize our work on this topic in the present section. We first present a brief derivation of an exact expression for the bicentric pdf in terms of a Fourier series. We shall then comment on the so-called partially bicentric pdfs, which are applicable to a multitude of real situations in which only part of the asymmetric unit of $P\bar{1}$ has a non-crystallographic centre of symmetry. We conclude this section with a discussion of the effect of several non-crystallographic centres of symmetry.

5.7.1 *The Fourier series for the bicentric pdf in* $P\bar{1}$

We recall the definition of a bicentric arrangement. The asymmetric unit of the centrosymmetric space group consists of two identical parts, which will be referred to as asymmetric subunits. These are related by a non-crystallographic centre of symmetry. If we assume that the non-crystallographic centre of symmetry as well as all the atoms are located in general Wyckoff positions of the space group, then in the case of $P\bar{1}$, the asymmetric subunit contains $N/4$ atoms, where N is the number of atoms in the unit cell. The structure factor for this arrangement is given by eqn(1.2.16) and the expression for the normalized structure factor is

$$E(\mathbf{h}) = 4\cos\delta \sum_{j=1}^{N/4} n_j \cos(\theta_j - \delta), \tag{5.7.75}$$

where $\delta = 2\pi\mathbf{h}^T\mathbf{d}$, $\mathbf{d}$ being the position vector of the non-crystallographic centre of symmetry, and $\theta_j = 2\pi\mathbf{h}^T\mathbf{r}_j$, where the $\mathbf{r}_j$s are the position vectors of the $N/4$ atoms within the asymmetric subunit. The characteristic function for the required pdf of E is given by

$$C(\omega_1) = \left\langle \exp\left[4i\omega_1 \sum_{j=1}^{N/4} n_j \cos\delta\cos(\theta_j - \delta)\right]\right\rangle_{\theta_1,\ldots,\theta_{N/4},\delta} \tag{5.7.76}$$

(cf. Shmueli, Weiss and Kiefer 1985). The averaging on the right-hand side of eqn(5.7.76) is performed first over the θs and finally over the angle δ. In this way we find

$$C(\omega_1) = \frac{1}{2\pi}\int_{-\pi}^{\pi} d\delta \prod_{j=1}^{N/4} \left[\frac{1}{2\pi}\int_{-\pi}^{\pi} \exp(4i\omega_1 n_j \cos\delta\cos(\theta-\delta)d\theta\right] \tag{5.7.77}$$

$$= \frac{2}{\pi}\int_0^{\pi/2} \left[\prod_{j=1}^{N/4} J_0(4\omega_1 n_j \cos\delta)\right] d\delta. \tag{5.7.78}$$

The Fourier representation of the pdf of $|E|$ for the space group $P\overline{1}$ is therefore given by

$$p(|E|) = \frac{1}{E_{max}}\left[1 + 2\sum_{m=1}^{\infty} C(\omega_m)\cos(\omega_m E)\right], \tag{5.7.79}$$

where the Fourier coefficient $C(\omega_m)$ is given by the value of the characteristic function in eqn(5.7.78) evaluated at $\omega_1 = \omega_m = \pi m/E_{max}$.

Figure 5.8 shows a comparison of the result just derived [eqn(5.7.79) (solid line) with the Fourier coefficients obtained from eqn(5.7.78)], the bicentric pdf based on the central limit theorem (dashed line), given by eqn(3.3.71), and a simulated pdf, based on a direct sampling of eqn(5.7.75). Figure 5.8*a* shows such a comparison for the equal-atom case, with 15 atoms in the asymmetric subunit. In this case all three pdfs agree quite well, as might have been expected, since the Fourier representation of the bicentric pdf reduces to the Rogers and Wilson (1953) pdf for a sufficiently large number of equal atoms (Shmueli, Weiss and Kiefer 1985; Appendix A). Figure 5.8*b* shows the comparison for an asymmetric subunit of composition $C_{14}U$. The discrepancy between the pdf based on the central limit theorem and its Fourier (or simulated) representation is now considerable. However, the discrepancy is much smaller than that displayed in Fig. 5.1. The reason for this may be the fact that there are now four rather than two uranium atoms in the unit cell, with a consequent decrease in the degree of atomic heterogeneity.

5.7.2 *Partially bicentric Fourier pdf of* $|E|$

A configuration which occurs rather frequently in real structures involves the presence of partial non-crystallographic centrosymmetry. For example, assume that 1,2-dimethylanthracene crystallizes in space group $P\overline{1}$ with two molecules in the unit cell. The two methyl substituents destroy the exact centrosymmetry of the asymmetric unit which, however, contains a large (14-atom) centrosymmetric portion — the anthracene fragment. The asymmetric unit is approximately centrosymmetric, leading one to expect that the appropriate pdf of $|E|$ for this crystal will be approximately bicentric. This problem was examined by Shmueli and Weiss (1985a) and merits a brief discussion.

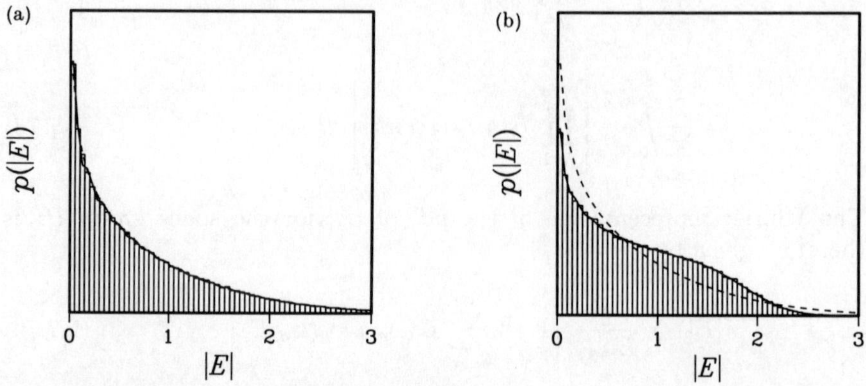

FIG. 5.8. *Ideal, exact and simulated bicentric pdfs.* (a) The equal-atom case. (b) The $C_{14}U$ asymmetric subunit.

The derivation of the partially bicentric pdf is a direct extension of the derivation given in the last subsection. Let the unit cell of $P\bar{1}$ contain $N = N_B + N_R$ atoms, where N_B of them constitute two centrosymmetric fragments containing non-crystallographic centres of symmetry. The remaining N_R atoms are related in pairs by the crystallographic centre of symmetry. All the atoms, as well as the non-crystallographic centre at $\mathbf{d}$, are assumed to occupy general Wyckoff positions. The normalized structure factor for this arrangement is found by combining the expressions for a purely bicentric and a 'simple' centrosymmetric arrangement of atoms in space group $P\bar{1}$

$$E(\mathbf{h}) = 4 \sum_{j=1}^{N_B/4} n_j \cos\delta \cos(\theta_j - \delta) + 2 \sum_{k=1}^{N_R/2} n_k \cos\psi_k, \tag{5.7.80}$$

where $\delta = 2\pi\mathbf{h}^T\mathbf{d}$, $\theta_j = 2\pi\mathbf{h}^T\mathbf{r}_{jB}$ and $\psi_k = 2\pi\mathbf{h}^T\mathbf{r}_{kR}$ are assumed to be independent random variables uniformly distributed on $(-\pi, \pi)$. The position vectors $\mathbf{r}_{jB}$ and $\mathbf{r}_{kR}$ belong to atoms within the centrosymmetric fragment of the asymmetric unit and outside it respectively. The characteristic function for the partially bicentric pdf is therefore

$$\begin{aligned} C(\omega_1) &= \langle \exp(i\omega E) \rangle \\ &= \left\langle \exp\left\{ i\omega_1 \left[4 \sum_{j=1}^{N_B/4} n_j \cos\delta \cos(\theta_j - \delta) + 2 \sum_{k=1}^{N_R/2} n_k \cos\psi_k \right] \right\} \right\rangle \end{aligned} \tag{5.7.81}$$

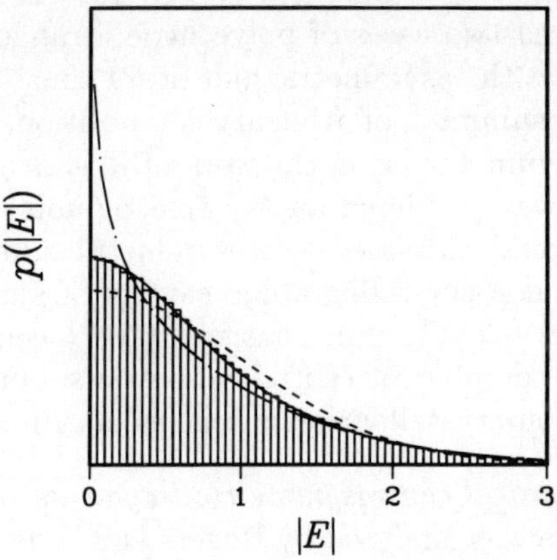

FIG. 5.9. *Ideal bicentric, exact partially bicentric, ideal centric and simulated pdfs.* Unit cell contents: C_{76}, $N_B = 60$, $N_R = 16$.

$$= \left\langle \exp\left[4i\omega_1 \sum_{j=1}^{N_B/4} n_j \cos\delta \cos(\theta_j - \delta)\right]\right\rangle$$

$$\times \left\langle \exp\left(2i\omega_1 \sum_{k=1}^{N_R/2} n_k \cos\psi_k\right)\right\rangle . \tag{5.7.82}$$

The averages in eqn(5.7.82) have already appeared in section 5.1 and subsection 5.6.1, and may be evaluated in the same manner as given there. The final result for the characteristic function is therefore equal to

$$C(\omega_1) = \left\{\frac{2}{\pi}\int_0^{\pi/2}\left[\prod_{j=1}^{N_B/4} J_0(4\omega_1 n_j \cos\delta)\right] d\delta\right\}\left\{\prod_{k=1}^{N_R/2} J_0(2\omega_1 n_k)\right\} \tag{5.7.83}$$

and the Fourier coefficient $C(\omega_m)$ is obtained by replacing ω_1 in eqn(5.7.83) by $\omega_m = \pi m/E_{max}$. The Fourier series in eqn(5.7.79) can be used to compute the pdf of $|E|$.

Figure 5.9 illustrates the ideal bicentric pdf (dashed-dotted line), the exact partially bicentric pdf (solid line), the ideal centric pdf (dashed line) and the corresponding simulated pdf. One sees that the partially bicentric pdf admits the exact centric and bicentric pdfs as special cases.

5.7.3 *Multicentric pdfs of $|E|$ in space group $P\overline{1}$*

The previous subsections dealt with effects of the presence of a single non-crystallographic centre of symmetry in the asymmetric unit of the space group. Several such centres may coexist (Rogers and Wilson 1953). Their effects have been analysed by Shmueli, Weiss and Wilson (1989).

We have addressed two cases of polycentric arrangements. In the first case, we assumed that the asymmetric unit of $P\bar{1}$ consists of several identical centrosymmetric subunits, of arbitrary composition, interrelated by additional centres of symmetry [as in the work of Rogers and Wilson (1953)]. In the second case, we considered an asymmetric unit of the space group, which consists of several *unrelated* centrosymmetric subunits. Thus in case 1, if the number of non-crystallographic centres of symmetry within the asymmetric unit is $n-1$, the unit consists of 2^{n-1} centrosymmetric subunits. In case 2, the number of centrosymmetric subunits is the same as the number of the non-crystallographic centres of symmetry.

5.7.3.1 *Case 1: Identical centrosymmetric fragments related by centres of symmetry* This case was analysed by Rogers and Wilson (1953) who used the central limit theorem, together with the tacit assumption of an equal-atom structure. In this chapter we allow the fundamental motif to have an arbitrary composition. An analysis based on the exact Fourier representation of the pdf leads to calculations that are formally quite simple. The normalized structure factor for the Rogers and Wilson multicentric arrangement is

$$E = 2^n \cos\psi_2 \cos\psi_3 \cdots \cos\psi_n \sum_{j=1}^{m/2} n_j \cos\theta_j, \tag{5.7.84}$$

where $\psi_k = 2\pi\mathbf{h}^T\mathbf{d}_k$ and $\theta_j = 2\pi\mathbf{h}^T\mathbf{r}_j$. In this equation, $\mathbf{d}_k$ is the position vector of the kth independent non-crystallographic centre of symmetry, n_j is the normalized scattering factor of the jth atom, and m is the number of atoms in the fundamental centrosymmetric fragment. The Fourier coefficient required for calculating the pdf is

$$C_m = \left(\frac{2}{\pi}\right)^{n-1} \int_0^{\pi/2} \cdots \int_0^{\pi/2} \left[\prod_{j=1}^{m/2} J_0\left(\frac{2^n \pi m n_j}{E_{max}} \prod_{k=2}^{n} \cos\psi_k\right)\right] \prod_{k=2}^{n} d\psi_k. \tag{5.7.85}$$

An approximate form of the above pdf for a structure composed of a large number of nearly equal atoms in the unit cell has been given by Rogers and Wilson in terms of the normalized intensity. In terms of the magnitude of the normalized structure factor their limiting pdf can be expressed as

$$p(|E|) = \left[\frac{2^n}{\pi^{(2n-1)}}\right]^{1/2} \int_0^{\pi/2} \cdots \int_0^{\pi/2} \exp\left[-\frac{E^2}{2^n}\sec^2\psi_2 \cdots \sec^2\psi_n\right] \times \sec\psi_2 \cdots \sec\psi_n d\psi_2 \cdots d\psi_n. \tag{5.7.86}$$

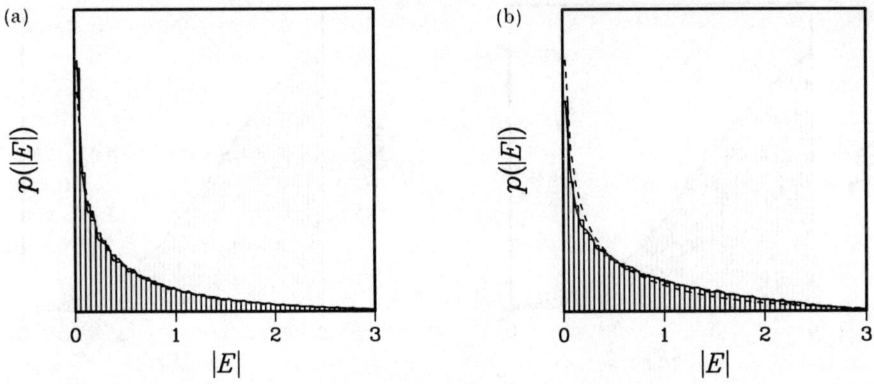

FIG. 5.10. *Ideal, exact and simulated tricentric pdfs (Case 1).* (a) The equal-atom case. (b) The $C_{14}U$ asymmetric subunit.

Figure 5.10 compares the ideal (solid line) tricentric ($n = 3$) pdf with Fourier representations (dashed line) of the tricentric pdf for the equal-atom case (Fig. 5.10*a*), and for a heterogeneous atomic composition (Fig. 5.10*b*). Both comparisons include simulated histograms based on eqn(5.7.84).

5.7.3.2 *Case 2: $n-1$ unrelated centrosymmetric fragments in the asymmetric unit* Let the ith centrosymmetric fragment in the asymmetric unit be centred at $\mathbf{d}_i$ and contain m_i atoms. Its contribution to the structure factor is

$$E(\mathbf{h}) = 4\cos\delta_i \sum_{j=1}^{m_i/4} n_j^{(i)} \cos(\theta_j - \delta_i), \qquad (5.7.87)$$

where $n_j^{(i)}$ is the normalized scattering factor of the jth atom in the ith centrosymmetric fragment. Since the fragments are assumed to be unrelated, the structure factor is simply the sum of the individual contributions

$$E(\mathbf{h}) = \sum_i^n E^{(i)}(\mathbf{h}) \qquad (5.7.88)$$

and the characteristic function for the corresponding pdf of E is the product of the characteristic functions for each individual fragment. These partial characteristic functions have the same functional form as that derived for the pure bicentric pdf [see eqn(5.7.76)]. Hence we can write

$$C(\omega_1) = \langle \exp(i\omega_1 E) \rangle$$

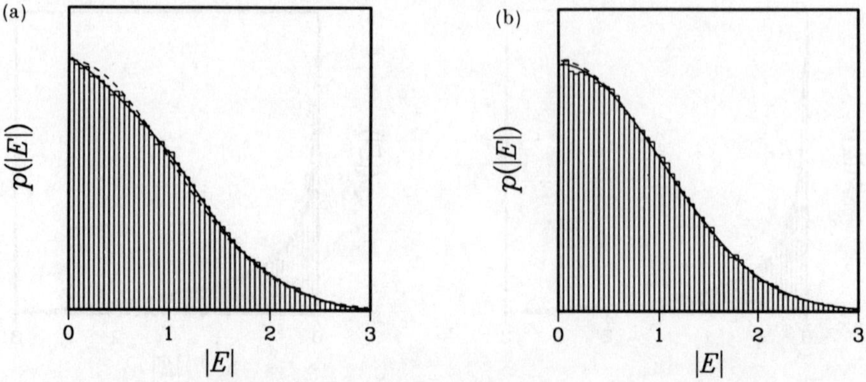

FIG. 5.11. *Exact and simulated tricentric pdfs of* $|E|$ *(Case 2).* (a) Two unrelated heterogenous fragments. (b) Five unrelated heterogeneous fragments.

$$= \prod_{i=2}^{n} C^{(i)}(\omega_1), \tag{5.7.89}$$

with

$$C^{(i)}(\omega_1) = \frac{2}{\pi} \int_0^{\pi/2} \left[\prod_{j=1}^{N/4} J_0(4\omega_1 n_j^{(i)} \cos\delta) \right] d\delta. \tag{5.7.90}$$

The Fourier coefficient to be used with eqn(5.7.79) is given by eqn(5.7.89), where the components are just these functions. We compare in Fig. 5.11 the Fourier pdf for several unrelated centrosymmetric fragments, with the corresponding simulated histograms, for two (Fig. 5.11*a*) and five (Fig. 5.11*b*) fragments, for a heterogeneous composition. It is evident that the pdf that describes intensity statistics for case 2 is quite close to the form predicted by the central limit theorem. This is rather encouraging since case 2 finds its counterparts in real structures more often than case 1 discussed earlier (e.g. in crystals of heteromolecular complexes composed of centrosymmetric molecules). It is also obvious that the relatively frequent bicentric arrangement gives rise to observable qualitative changes of the pdf for the structure-factor statistics and is perhaps the most important of all non-crystallographic symmetries, at least in the context of intensity statistics.

5.8 Unequally bounded components of the structure factor: the subcentric pdf, effects of dispersion and more about special positions

Our derivations of the pdfs of $|E|$ for non-centrosymmetric space groups were based on the assumption that the real and imaginary parts of E have common upper and lower bounds, i.e. if $E = A+iB$ we have always assumed that

$$-E_{max} \leq A, B \leq E_{max}$$

where

$$E_{max} = A_{max} = B_{max} = \sum_{j=1}^{N} n_j.$$

Our derivations commenced with first finding a joint pdf of A and B. As a second step A and B were represented in polar coordinate form as $|E| \cos\varphi$ and $|E| \sin\varphi$, where φ is the phase of the structure factor. The final step required an integration of the joint pdf with respect to φ over the interval $(-\pi,\pi)$. This is no longer possible when the maximum values of A and B are unequal. In fact, when this occurs the possible values of φ are restricted and the calculation of the pdfs becomes somewhat more complicated than in the case when A and B have common bounds. The general problem of finding Fourier representations of pdfs of $|E|$ for unequally bounded A and B was treated by Shmueli, Rabinovich and Weiss (1990). We follow that analysis in our application.

We shall discuss three examples of situations in which the problem of restricted phases is bound to appear:

1. The structure contains significantly dispersive scatterers so that the imaginary parts of some atomic scattering factors cannot be neglected. Therefore, even when the space group is centrosymmetric with the origin chosen at a centre of symmetry, the imaginary part, B, of the structure factor does not vanish. Its maximum value is generally much smaller than that of A [see tables of scattering factors in Volume C of *International Tables for Crystallography* (1992)] and the phase is necessarily restricted. Effects of dispersion on the pdf of the magnitude of the structure factor were first studied by Wilson (1980) using the central limit theorem (see section 3.4). Wilson's result was further generalized by Shmueli and Wilson (1983), who proposed appropriate correction-factor expansions of the Gram–Charlier type. In that treatment the problem of unequally bounded A and B never arose, since the magnitudes of A and B are unrestricted in approximations based on the central limit theorem.
2. The structure belongs to a non-centrosymmetric space group but the asymmetric unit contains one or more centrosymmetric structural fragments. The presence of such fragments ensures that the maximum

value of the real part of the structure factor exceeds the maximum value of its imaginary part. This implies a phase restriction that must be accounted for in an exact analysis of the problem. This kind of non-crystallographic symmetry was studied by a number of investigators (see Srinivasan and Parthasarathy, 1976 and references quoted therein; as well as Wilson 1980) in the central limit theorem approximation (see also section 3.4), and was further analysed by Shmueli and Wilson (1983), who also termed the corresponding distribution 'subcentric'.

3. The problem of calculating the pdf of $|E|$ allowing for the presence of scatterers in special positions has been investigated by Shmueli and Weiss (1988) (see also section 5.5). While most low-symmetry space groups can be treated on the assumption that the phase of the structure factor is uniform on $(-\pi, \pi)$, this is not possible for space groups isomorphous to the point group 222, since the imaginary parts of the trigonometric structure factors corresponding to special positions vanish identically by symmetry. The phase is then clearly restricted.

We shall first treat the general problem of deriving the pdf of $|E|$ assuming that the maximum values of A and B differ. In what follows we apply the results of this derivation to examples illustrative of the three cases just described.

5.8.1 *Pdf of $|E|$ when A and B are unequally bounded*

As in the other sections of this chapter, we first derive the joint pdf of A and B in terms of a Fourier series. Such an expansion is valid, since both A and B are bounded, even though the bounds may differ. We assume without loss of generality that $A_{max} > B_{max}$. The exact contour in the (A, B) complex plane surrounding the (A, B) values for which the joint pdf, $p(A, B)$, is non-zero is confined to the rectangle defined by the vertices $(-A_{max}, -B_{max})$, $(-A_{max}, B_{max})$, $(A_{max}, -B_{max})$ and (A_{max}, B_{max}), as shown in Fig. 5.12. The pdf $p(A, B)$ vanishes by definition at any point outside the exact contour — whether inside the indicated rectangle or outside it. The joint pdf can therefore be expanded in a two-dimensional Fourier series independently of the shape of the exact contour. The general form of such a series is

$$p(A, B) = \frac{1}{4A_{max}B_{max}} \sum_u \sum_v C_{uv} \exp\left[-\pi i \left(\frac{uA}{A_{max}} + \frac{vB}{B_{max}}\right)\right], \tag{5.8.91}$$

where

$$C_{uv} = \int_{-A_{max}}^{A_{max}} \int_{-B_{max}}^{B_{max}} p(A, B) \exp\left[\pi i \left(\frac{uA}{A_{max}} + \frac{vB}{B_{max}}\right)\right] dA dB$$

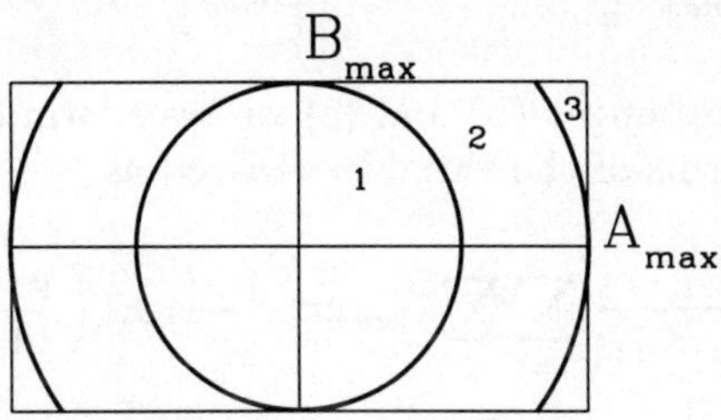

FIG. 5.12. *Locus of $E = A + iB$ for unequally bounded A and B.* The numbers 1, 2 and 3 correspond to the ranges (a), (b) and (c) in the text respectively.

$$\tag{5.8.92}$$

$$= \left\langle \exp\left[\pi i \left(\frac{uA}{A_{max}} + \frac{vB}{B_{max}}\right)\right]\right\rangle. \tag{5.8.93}$$

Indeed, the limits of integration in eqn(5.8.92) can be taken to coincide with the exact contour within which $p(A, B)$ may differ from zero. However, since this contour is included in the indicated rectangle, and there is no contribution from points outside the exact contour and within the rectangle, the simpler integration over the rectangle is permissible.

In order to obtain the required expressions for $p(|E|)$ we first transform (A,B) to the polar coordinates $(|E|,\varphi)$ and use the fact that

$$p(|E|, \varphi) = |E| p(|E| \cos\varphi, |E| \sin\varphi).$$

We then integrate out the phase, while taking account of the proper limits of integration. As shown in Fig. 5.12, three ranges of $|E|$ have to be considered in order to cover the whole rectangle described above:

(a) $|E| \leq B_{max} < A_{max}$
In this range the phase of E is allowed to take any value between $-\pi$ and π.
(b) $B_{max} < |E| \leq A_{max}$
The phase is restricted in this case and its limits, in the positive quadrant, are $\varphi_1 = 0$ and $\varphi_2 = \sin^{-1}(B_{max}/|E|)$.
(c) $A_{max} < |E| \leq (A_{max}^2 + B_{max}^2)^{1/2}$
The phase is again restricted and its limits, in the positive quadrant are: $\varphi_1 = \cos^{-1}(A_{max}/|E|)$ and $\varphi_2 = \sin^{-1}(B_{max}/|E|)$.

In range (a) the pdf of $|E|$ is analogous to that given by Shmueli and Weiss (1987) [see also eqn(5.2.19)],

$$p(|E|) = \frac{\pi|E|}{2B_{max}^2}\sum_u\sum_v C_{uv}J_0\left[\pi\frac{|E|}{B_{max}}(u^2+v^2)^{1/2}\right]. \qquad (5.8.94)$$

In order to treat the ranges (b) and (c) we transform eqn(5.8.91) so that it is expressed in terms of the variables $(|E|, \varphi)$ as

$$\begin{aligned} p(|E|,\varphi) &= \frac{|E|}{4A_{max}B_{max}}\sum_u\sum_v C_{uv}\exp\left[-\pi i|E|\left(\frac{u\cos\varphi}{A_{max}}+\frac{v\sin\varphi}{B_{max}}\right)\right] \\ &\equiv \frac{|E|}{4A_{max}B_{max}}\sum_u\sum_v C_{uv}\exp[-i(G_u\cos\varphi+H_v\sin\varphi)] \qquad (5.8.95) \end{aligned}$$

where $G_u = \pi|E|u/A_{max}$ and $H_v = \pi|E|v/B_{max}$. In order to integrate out the phase, we first sum the complex exponential in eqn(5.8.95) at the points φ, $-\varphi$, $\pi-\varphi$ and $\pi+\varphi$ and integrate over φ in the positive quadrant only. The pdf of $|E|$ then takes the form

$$p(|E|) = \frac{|E|}{A_{max}B_{max}}\sum_u\sum_v C_{uv}\int_{\varphi_1}^{\varphi_2}\cos(G_u\cos\varphi)\cos(H_v\sin\varphi)d\varphi. \qquad (5.8.96)$$

The integral in eqn(5.8.96) can be evaluated numerically, or can be represented in terms of series of Bessel functions after the integrand is expanded and integrated. For numerical integration, an adaptive method, such as the Romberg method (Davis and Rabinovitz 1967), is recommended. The alternative expansion of the integrand in a serics of Bessel functions is based on the equations

$$\cos(z\cos\varphi) = J_0(z) + 2\sum_{k=1}^{\infty}(-1)^k J_{2k}(z)\cos(2k\varphi) \qquad (5.8.97)$$

and

$$\cos(z\sin\varphi) = J_0(z) + 2\sum_{k=1}^{\infty} J_{2k}(z)\cos(2k\varphi) \qquad (5.8.98)$$

(Abramowitz and Stegun 1972). The final result for $p(|E|)$ can be expressed as

$$p(|E|) = \frac{|E|}{A_{max}B_{max}}[(\varphi_2-\varphi_1)+2(S_1+S_2+S_{12})], \qquad (5.8.99)$$

where

$$S_1 = \sum_{u=1}^{\infty} C_{u0}$$

$$\times \left\{ (\varphi_2 - \varphi_1) J_0(G_u) + \sum_{p=1}^{\infty} (-1)^p J_{2p}(G_u) K_p \right\}, \quad (5.8.100)$$

$$S_2 = \sum_{v=1}^{\infty} C_{0v}$$
$$\times \left\{ (\varphi_2 - \varphi_1) J_0(H_v) + \sum_{p=1}^{\infty} J_{2p}(H_v) K_p \right\} \quad (5.8.101)$$

and

$$S_{12} = \sum_{u=-\infty}^{\infty} \sum_{v=1}^{\infty} C_{uv}$$
$$\times \left\{ (\varphi_2 - \varphi_1) J_0(\zeta) + \sum_{p=1}^{\infty} (-1)^p J_{2p}(\zeta) K_p \cos(2p\Delta) \right\} \quad (5.8.102)$$

with

$$\zeta = (G_u^2 + H_v^2)^{1/2}, \quad (5.8.103)$$

$$K_p = \frac{\sin(2p\varphi_2) - \sin(2p\varphi_1)}{p} \quad (5.8.104)$$

and

$$\Delta = \tan^{-1}\left(\frac{vA_{max}}{uB_{max}}\right). \quad (5.8.105)$$

Notice that the summations in these equations extend over positive indices only, a simplification which follows from symmetry considerations. The double summation for S_{12} needs to be evaluated only for $u \geq v$, provided suitable internal symmetrization is performed (Shmueli, Rabinovich and Weiss 1989*a*; see also section 6.4).

5.8.2 *Derivation of the Fourier coefficients and illustrative examples*

We shall now derive the Fourier coefficients for the pdfs appropriate to the three situations outlined at the beginning of this section, and compare these pdfs with simulated ones. We shall show that the Fourier representation — besides being an exact one — is also appreciably simpler than the Gaussian approximation together with the corrections briefly discussed in subsection 4.3.2.

5.8.2.1 *Anomalous dispersion in space group $P\bar{1}$* The normalized structure factor can now be written as

$$E = 2\sum_{j=1}^{N/2}(n'_j + in''_j)\cos\theta_j, \tag{5.8.106}$$

and the real and imaginary parts of the normalized scattering factor are defined in eqn(3.4.75). The maximum values of A and B in this case are

$$A_{max} = \sum_{j=1}^{N/2} n'_j \text{ and } B_{max} = \sum_{j=1}^{N/2} n''_j \tag{5.8.107}$$

and the locus of E corresponding to this case is shown in Fig. 5.12. The characteristic function corresponding to the pdf of $|E|$ is given by

$$C(\omega_1, \omega_2) = \prod_{j=1}^{N/2} C_j(\omega_1, \omega_2), \tag{5.8.108}$$

where

$$C_j(\omega_1, \omega_2) = \frac{1}{2\pi}\int_{-\pi}^{\pi} \exp[2i(\omega_1 n'_j + \omega_2 n''_j)\cos\theta]d\theta \tag{5.8.109}$$

$$= J_0[2(\omega_1 n'_j + \omega_2 n''_j)]. \tag{5.8.110}$$

The Fourier coefficients appearing in eqns (5.8.94) and (5.8.96) are therefore given by

$$C_{uv} = \prod_{j=1}^{N/2} J_0\left[2\pi\left(n'_j\frac{u}{A_{max}} + n''_j\frac{v}{B_{max}}\right)\right]. \tag{5.8.111}$$

which are to be used with eqns (5.8.99) – (5.8.105). In Fig. 5.13 we compare the pdf of $|E|$ for a $P\bar{1}$ structure with asymmetric unit $C_{19}Pt$, allowing for anomalous dispersion (solid line), a pdf of $|E|$ for the same composition and symmetry in which anomalous dispersion is neglected (dashed line), and a simulated histogram based on 100 000 values of $|E|$. The effect of dispersion is significant only at very low values of $|E|$, and manifests itself in a small peak in the pdf. In contrast to the dispersionless case the pdf is equal to zero for $|E| = 0$ and increases steeply in the neighbourhood of the origin. Since these low-magnitude $|E|$ values usually correspond to very weak reflections, the effect may be difficult to detect unless the intensities of weak and 'unobserved' reflections are processed in a very careful manner.

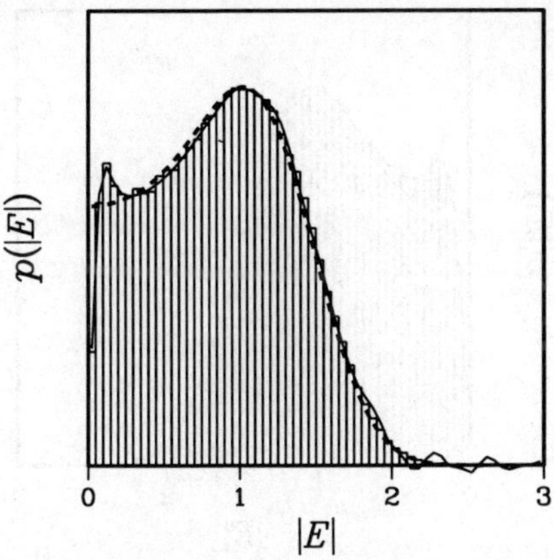

FIG. 5.13. *Exact pdfs of $|E|$ in $P\bar{1}$ allowing for dispersion and neglecting it.* The histogram accounts for the presence of dispersion.

5.8.2.2 *A subcentric arrangement in space group $P1$* An exact characteristic function for this example was briefly discussed elsewhere (Weiss *et al.* 1985), a more detailed derivation later being given by Shmueli, Rabinovich and Weiss (1990). Let the unit cell contain N atoms and let M of them constitute a centrosymmetric fragment. The remaining $N - M$ atoms in the unit cell are not related by any symmetry operation.

If the origin of the space group $P1$ is chosen at the (non-crystallographic) centre of symmetry the normalized structure factor is given by

$$E = 2\sum_{j=1}^{M/2} n_j \cos\theta_j + \sum_{k=M+1}^{N} n_k(\cos\theta_k + i\sin\theta_k) \tag{5.8.112}$$

$$\equiv A + iB. \tag{5.8.113}$$

The maximum values of A and B are, in the present case,

$$A_{max} = 2\sum_{j=1}^{M/2} n_j + \sum_{k=M+1}^{N} n_k \tag{5.8.114}$$

and

$$B_{max} = \sum_{k=M+1}^{N} n_k, \tag{5.8.115}$$

which can be used to construct a Fourier representation of the subcentric pdf of $|E|$. The locus of E is also contained in the rectangle shown in

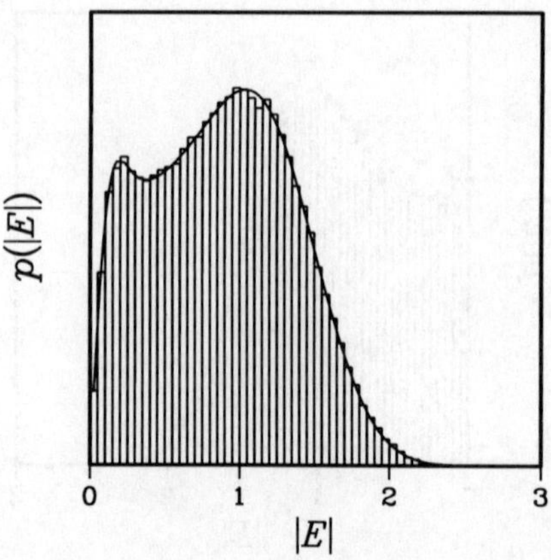

FIG. 5.14. *Exact and simulated pdf of $|E|$ for $P1$ allowing for a subcentric arrangement, atomic heterogeneity and effects of dispersion.*

Fig. 5.12. The Fourier coefficients for the pdf in question are obtained by evaluating the average in eqn(5.8.93), with A and B defined as in eqns (5.8.113) and (5.8.112). The result is

$$C_{uv} = \prod_{j=1}^{M/2} J_0\left(\frac{2\pi u n_j}{A_{max}}\right) \prod_{k=M+1}^{N} J_0\left[\pi n_k \left(\frac{u^2}{A_{max}^2} + \frac{v^2}{B_{max}^2}\right)^{1/2}\right], \tag{5.8.116}$$

where dispersion has been neglected. If we want to account for dispersion in the centrosymmetric fragment, all we have to do is to replace un_j/A_{max} in the first product in eqn(5.8.116) by $un'_j/A_{max}+vn''_j/B_{max}$ for the dispersive scatterer. The pdf is then computed from eqns (5.8.94) or (5.8.96) according to the value of E defined by eqn(5.8.112).

Figures 5.14 and 5.15 show some pdfs of $|E|$ for the subcentric arrangement. Figure 5.14 shows the pdf for a $C_{43}Pt_2$ unit comprising a centrosymmetric fragment with composition $C_{38}Pt_2$, and five carbon atoms unrelated by symmetry. The dispersion of the platinum atom has been taken into account in calculating the pdf. As might be expected, there is only a slight departure from a centrosymmetric dispersionless pdf, mainly in the region of small $|E|$ values. Figure 5.15 corresponds to a subcentric arrangement in the equal-atom case. All three curves correspond to the case of 55 atoms in the unit cell, subdivided into a centrosymmetric fragment (C) and unrelated atoms (N). It is interesting to note that even the extreme pseudo-centrosymmetry in Fig. 5.15a leads to a pdf significantly different from the ideal centric one.

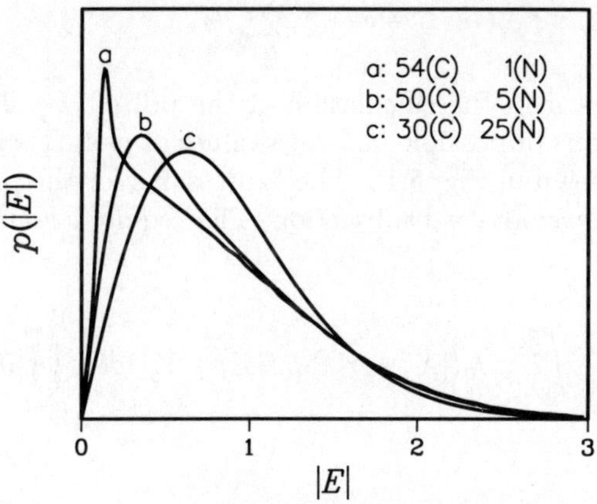

FIG. 5.15. *Exact pdfs of* $|E|$ *for equal-atom subcentric arrangements in* $P1$.

5.8.2.3 *Special positions in space group* $P222$ Let the asymmetric unit of the space group $P222$ contain M atoms, M_g of them occupying general positions and $M_{sp} = M - M_g$ being located on twofold axes of the space group. The Fourier coefficient for the pdf corresponding to this arrangement is a product of contributions from atoms in general positions and from those lying in the special ones. The contributions of atoms lying in general positions can be taken from Shmueli and Weiss (1987), while those from the special positions are obtained by noting that the trigonometric structure factor for an atom located on a twofold axis of $P222$ has the form

$$\xi_s = 2\cos\theta \text{ and } \eta_s = 0 \tag{5.8.117}$$

for any of the three different axes of rotation. The structure factor for the indicated arrangement is therefore given by

$$E = \sum_{j=1}^{M_g} n_j(\xi_{gj} + i\eta_{gj}) + \sum_{k=1}^{M_{sp}} n_k \xi_{sj}, \tag{5.8.118}$$

where $\xi_{gj} + i\eta_{gj}$ is the trigonometric structure factor for an atom located in a general position of the space group. The maximum values of the real and imaginary parts of E are now given by

$$A_{max} = \sum_{j=1}^{M_g} n_j + \sum_{k=1}^{M_{sp}} n_k \tag{5.8.119}$$

and

$$B_{max} = \sum_{j=1}^{M_g} n_j, \tag{5.8.120}$$

which will be used in the calculation of the pdf of $|E|$. The contour surrounding the locus of possible non-zero values of $p(A, B)$ clearly lies within the rectangle shown in Fig. 5.12. The expressions for the pdf are analogous to those in the previous subsubsection. The required Fourier coefficient is given by

$$C_{uv} = \prod_{j=1}^{M_g} \left[\frac{2}{\pi} \int_0^{\pi/2} J_0(|X_j + Y_j|) J_0(|X_j - Y_j|) d\theta \right] \prod_{k=1}^{M_{sp}} J_0 \left(\frac{2\pi u n_k}{A_{max}} \right), \tag{5.8.121}$$

where

$$X_j = \frac{2\pi u n_j}{A_{max}} \cos\theta \tag{5.8.122}$$

$$Y_j = \frac{2\pi v n_j}{B_{max}} \sin\theta. \tag{5.8.123}$$

Figure 5.16 illustrates the results just derived. The specific examples are those of hypothetical crystals in space group $P222$, in one of which the asymmetric unit contains $C_{19}Pt$, all atoms being located in general positions of the space group, while in the other the general positions are occupied in the same manner as in the first with an additional Pt atom located on a twofold axis of $P222$ (Fig. 5.16). These two pdfs are compared with the ideal acentric pdf (Wilson 1949). All pdfs have been compared with simulated histograms (not shown). The three pdfs are remarkably similar, which shows that intensity statistics for this space group are relatively insensitive to atomic heterogeneity and to the presence of heavy scatterers in special positions. If, however, the general positions were occupied by light atoms only, and a heavy atom were located on a twofold axis, it is probable that the pdf would be similar to that found for space group $P\bar{1}$ type.

All the calculations so far illustrated in this chapter include quantitative comparisons with the simulated histograms by means of the corresponding discrepancy or R factors, defined as

$$R = \left\{ \frac{\sum_i [p_{calc}^{(i)} - p_{sim}^{(i)}]^2}{\sum_i [p_{sim}^{(i)}]^2} \right\}^{1/2}, \tag{5.8.124}$$

where $p_{sim}^{(i)}$ is the height of the rectangle of the ith bin, and $p_{calc}^{(i)}$ is the value of the theoretical pdf calculated at the midpoint of the ith bin. The theoretical pdfs are multiplied by scale factors which make them comparable

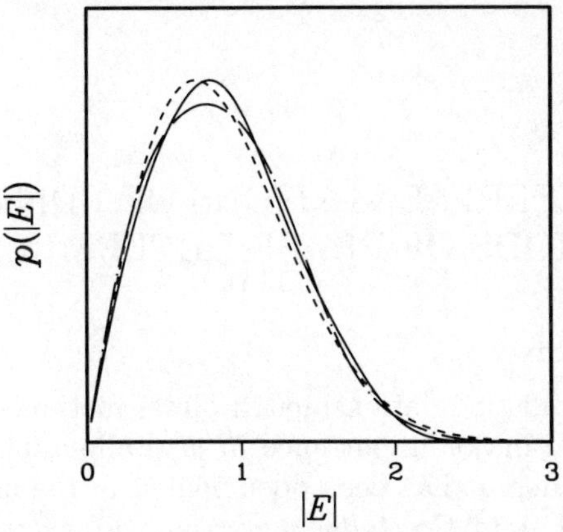

FIG. 5.16. *Exact and ideal pdfs of $|E|$ for the space group* $P222$. All the heavy atoms in general positions (solid), heavy atoms both in general and special positions (dashed-dotted) and ideal acentric (dashed).

with the (unscaled) simulated histograms. Good visual agreement between pdfs is associated with R values of 0.01 to 0.03.

6

FOURIER REPRESENTATIONS OF PDFS IN DIRECT METHODS OF PHASE DETERMINATION

6.1 Introduction

The role of probabilistic relationships in direct methods of phase determination has been of major importance in crystallographic practice during the past four decades. It has been documented in the crystallographic literature, mainly in *Acta Crystallographica* and in a number of books and monographs. The extension from pdfs of the magnitude of a single structure factor, of importance in intensity statistics, to the joint pdfs of several structure factors, on which direct methods of sign and phase determination are based was first discussed in the well-known monograph of Hauptman and Karle (1953*b*). A number of joint pdfs are further developed in the papers of Vand and Pepinsky (1953), Woolfson (1954), Cochran and Woolfson (1955) and Cochran (1955), and in many subsequent publications by Karle and Hauptman. An early and very clear exposition of corrections to Gaussian approximations of joint pdfs of several centrosymmetric structure factors was presented by Klug (1958) and both the centrosymmetric and non-centrosymmetric cases, using the Edgeworth expansion, are treated in two detailed papers by Naya, Nitta and Oda (1964,1965). Gaussian approximations to the joint pdfs of three structure factors were summarized by Woolfson (1961) and Karle and Karle (1966), who also proposed multisolution techniques that rendered the solution of the phase problem for crystals of small and medium-sized molecules a practical proposition. A detailed survey of the literature on direct methods is beyond the scope of this monograph. Extensive summaries and lists of references to this literature is given in the book of Giacovazzo (1980), the collection articles by Ladd and Palmer (1980), Proceedings of Schools on Direct Methods (e.g. Schenk, 1990) and in a review article by Giacovazzo (1993). Further relevant material is scattered throughout *Acta Crystallographica* (Section A) and the proceedings of all recent international and regional crystallographic meetings.

The probabilistic techniques employed in all the voluminous literature cited above can be briefly classified as

1. Those based on the central limit theorem (e.g. Cochran 1955; see also section 3.5).

2. Correction to the central limit theorem based on Edgeworth or Gram–Charlier expansions of the joint pdfs of several structure factors (e.g. Peschar and Schenk 1987).
3. Maximum-entropy methods (e.g. Bricogne 1984).
4. Fourier representations of the joint pdfs of several structure factors (e.g. Shmueli and Weiss 1985b).

We conclude this monograph with a chapter describing our approach to the application of exact Fourier representations of joint pdfs to several relationships which are of importance in direct methods, and indicate how these exact probability measures may be applied in practice. The next section will deal with the development of a Fourier conditional pdf for the simplest sign relationship, the so-called Σ_1 formula. We shall then describe the conditional probability for the positive sign of the triple product $E_{\mathbf{h}}E_{\mathbf{k}}E_{-\mathbf{h}-\mathbf{k}}$ in space group $P\bar{1}$ and discuss exact Fourier representations of conditional pdfs for general three-phase invariants. We shall conclude the chapter with our recent work on a simple and accurate parametric approximation to the pdf of the three-phase invariant, based on the evaluation of the exact but numerically demanding expressions.

6.2 Exact Fourier representation of the probability for the Σ_1 relationship in space group $P\bar{1}$

When the magnitudes of the unitary structure factors of two reflections with diffraction vectors $\mathbf{h}$ and $2\mathbf{h}$ exceed a certain threshold, the sign of the structure factor of the $2\mathbf{h}$ reflection is determined to be positive by the simplest Harker–Kasper inequality

$$U_{\mathbf{h}}^2 \leq \frac{1}{2}(1 + U_{2\mathbf{h}}) \tag{6.2.1}$$

(Harker and Kasper 1948). In practice, however, the magnitudes of the structure factors are very seldom large enough for this inequality to be directly applicable. Yet, a set of intensity data collected from a crystal which belongs to space group $P\bar{1}$ may contain a number of pairs of reflections corresponding to diffraction vectors $\mathbf{h}$ and $2\mathbf{h}$, and associated with rather large structure factors — even if not large enough for the Harker–Kasper inequality. In the absence of a deterministic approach, one can ask for the probability that $U_{2\mathbf{h}}$ is positive, given its magnitude and the magnitude of $U_{\mathbf{h}}$. This probabilistic version of the Harker–Kasper inequality given by eqn(6.2.1) is known as the Σ_1 relationship, symbolically denoted as $p_+(2\mathbf{h}|\mathbf{h})$, i.e the probability that the sign of $U_{2\mathbf{h}}$ is positive conditioned on the knowledge of $|U_{\mathbf{h}}|$ and $|U_{2\mathbf{h}}|$. This conditional probability was investigated in the earliest days of phase determination by methods based on the central limit theorem. The expression for this probability — not unfrequently applied or consulted during the process of phase determination —

is based on the study of Cochran and Woolfson (1955) and, represented in terms of normalized structure factors, is given by

$$p_+(2\mathbf{h}|\mathbf{h}) \simeq \frac{1}{2} + \frac{1}{2}\tanh\left[\frac{\sigma_3}{2\sigma_2^{3/2}}K\right], \tag{6.2.2}$$

where $K = |E(2\mathbf{h})|(|E(\mathbf{h})|^2 - 1)$ and $\sigma_n = \sum_{j=1}^{N} Z_j^n$. Since the central limit theorem was used to derive eqn(6.2.2) it is only the lowest-order term in a sequence of approximations. We therefore thought it both useful and interesting to derive an exact expresion for this conditional probability in order to examine the dependence of $p_+(2\mathbf{h}|\mathbf{h})$ on the following factors:

1. The magnitudes of $|E_{\mathbf{h}}|$ and $|E_{2\mathbf{h}}|$.
2. The number of atoms in the asymmetric unit.
3. Atomic heterogeneity.

A derivation of the Fourier representation of the probability $p(2\mathbf{h}|\mathbf{h})$ as well as analysis of its dependence on these factors is contained in an article by Shmueli and Weiss (1985b). We follow this reference in the present discussion.

6.2.1 *Derivation*

6.2.1.1 *Mathematical analysis* The technique of Fourier representation of pdfs of a single structure factor, as developed in the last chapter, is readily generalized to yield formal representations for joint pdfs of several structure factors.

Let $\mathbf{E} = (E_{\mathbf{h}_1}, E_{\mathbf{h}_2}, \ldots, E_{\mathbf{h}_d})$ be the vector of normalized structure factors, where

$$E_{\mathbf{h}_s} = \sum_{j=1}^{N/g} n_j \xi_j(\mathbf{h}_s), \tag{6.2.3}$$

n_j is the normalized scattering factor and

$$\xi_j(\mathbf{h}_s) = \sum_{q=1}^{g} \cos[2\pi \mathbf{h}_s^T(\mathbf{P}_q\mathbf{r}_j + \mathbf{t}_q)] \tag{6.2.4}$$

is the real part of the trigonometric structure factor of the jth atom. In eqn(6.2.3) N is the number of atoms in the unit cell, and in eqn(6.2.4) $\mathbf{h}_s$ is the diffraction vector corresponding to the normalized structure factor $E_{\mathbf{h}_s}$. In eqn(6.2.4), g is the order of the point group times the multiplicity of the Bravais lattice and $(\mathbf{P}_q, \mathbf{t}_q)$ is the qth space-group operator.

The following analysis is confined to the centrosymmetric case and is based on the following premises:

1. The atomic phase factors, $2\pi\mathbf{h}^T\mathbf{r}$, are uniformly distributed on $(-\pi,\pi)$, and position vectors of different atoms are rationally independent. This implies that all the atoms are located in general Wyckoff positions.
2. The contributions of the different atoms within the asymmetric unit to the structure factor are statistically independent.
3. Effects of dispersion and non-crystallographic symmetry are neglected.

We derive the joint pdf of the components of the vector $\mathbf{E}$. Clearly, each individual component, say $E_{\mathbf{h}_s}$, can differ from zero only in the range $-E_{max} \le E_{\mathbf{h}_s} \le E_{max}$, where

$$E_{max} = \sum_{j=1}^{N} n_j \equiv 1/\alpha. \tag{6.2.5}$$

The required joint pdf can therefore be expanded in the multiple Fourier series

$$p(\mathbf{E}) = \left(\frac{\alpha}{2}\right)^d \sum_{\mathbf{u}} C_{\mathbf{u}} \exp(-\pi i\alpha\mathbf{u}^T\mathbf{E}), \tag{6.2.6}$$

where $\mathbf{u} = (u_1, u_2, \ldots, u_d)$ is a vector of the Fourier summation indices. The coefficients, $C_{\mathbf{u}}$, are found from

$$C_{\mathbf{u}} = \int_{-1/\alpha}^{1/\alpha} \cdots \int_{-1/\alpha}^{1/\alpha} p(\mathbf{E}) \exp(\pi i\alpha\mathbf{u}^T\mathbf{E}) d^m\mathbf{E} \tag{6.2.7}$$

$$= \int_{-\infty}^{\infty} \cdots \int_{-\infty}^{\infty} p(\mathbf{E}) \exp(\pi i\alpha\mathbf{u}^T\mathbf{E}) d^m\mathbf{E} \tag{6.2.8}$$

$$= C(\pi\alpha\mathbf{u}), \tag{6.2.9}$$

where m is the number of independent component structure factors, $E_{\mathbf{h}_s}$, and $C(\pi\alpha\mathbf{u})$ is the characteristic function corresponding to $p(\mathbf{E})$ evaluated at $\boldsymbol{\omega} = \pi\alpha\mathbf{u}$. For example, if $p(\mathbf{E}) = p(E_{\mathbf{h}}, E_{2\mathbf{h}})$, a double Fourier series as in eqn(6.2.6) is needed for the pdf, while only one integral needs to be evaluated in eqn(6.2.7), because $E_{\mathbf{h}}$ and $E_{2\mathbf{h}}$ are correlated. The details of this relation are given in the next subsection.

The characteristic function for the pdf in eqn(6.2.6), needed for the calculation of the Fourier coefficients $C_{\mathbf{h}}$, is expressible as

$$C(\boldsymbol{\omega}) = \langle \exp(i\boldsymbol{\omega}^T\mathbf{E}) \rangle \tag{6.2.10}$$

$$= \left\langle \exp\left(i\sum_{r=1}^{d} \omega_r E_{\mathbf{h}_r}\right)\right\rangle \tag{6.2.11}$$

$$= \left\langle \exp\left[i \sum_{r=1}^{d} \sum_{j=1}^{N/g} \omega_r n_j \xi_j(\mathbf{h}_r) \right] \right\rangle \tag{6.2.12}$$

$$= \prod_{j=1}^{N/g} \left\langle \exp\left[i n_j \sum_{r=1}^{d} \omega_r \xi_j(\mathbf{h}_r) \right] \right\rangle \tag{6.2.13}$$

$$= \prod_{j=1}^{N/g} C_j(\boldsymbol{\omega}). \tag{6.2.14}$$

where $C_j(\boldsymbol{\omega})$ is the contribution of the jth atom to the multivariate characteristic function, $C(\boldsymbol{\omega})$. We first expand the scalar product in eqn(6.2.10) to obtain eqn(6.2.11), in terms of individual normalized structure factors. We then substitute for $E_{\mathbf{h}_r}$ from eqn(6.2.3) and express the characteristic function in terms of atomic composition and trigonometric structure factors (i.e. crystal symmetry) in eqn(6.2.12). If we now make use of the assumed independence of atomic contributions, eqn(6.2.12) can be rewritten as the product of averages in eqn(6.2.13). The construction of the joint pdf (6.2.6) thus requires the evaluation of the atomic contributions, $C_j(\boldsymbol{\omega})$, and a replacement of the components of $\boldsymbol{\omega}$ by the appropriate functions of summation indices. In the present case, $\omega_r \to \pi \alpha u_r$.

The formalism just introduced is a general one for all centrosymmetric space groups, and can be modified to include all of the cases of non-crystallographic symmetry and dispersion. The corresponding treatment of non-centrosymmetric space groups is similar to that of the centrosymmetric ones.

6.2.1.2 *The conditional probability for the Σ_1 relationship in $P\bar{1}$* The joint pdf required in this example is $p(E_{\mathbf{h}}, E_{2\mathbf{h}})$, and this will lead to an exact expression for the probability that $E_{2\mathbf{h}}$ is positive, given the known magnitudes $|E_{2\mathbf{h}}|$ and $|E_{\mathbf{h}}|$.

We have, for the space group $P\bar{1}$,

$$\xi_j(\mathbf{h}) = 2\cos\theta_j \text{ and } \xi_j(2\mathbf{h}) = 2\cos 2\theta_j, \tag{6.2.15}$$

The atomic characteristic function, $C(\omega_1, \omega_2)$, is therefore obtained from eqns (6.2.13) and (6.2.15) in the form

$$C_j(\omega_1, \omega_2) = \langle \exp[2in_j(\omega_1 \cos\theta_j + \omega_2 \cos 2\theta_j)] \rangle \tag{6.2.16}$$

$$= \frac{1}{2\pi} \int_{-\pi}^{\pi} \exp[2in_j(\omega_1 \cos\theta + \omega_2 \cos 2\theta)] d\theta. \tag{6.2.17}$$

This integral can be solved numerically, and is also conveniently reducible to a rapidly converging double series. This is done by making use of the identity

$$\exp(iu\cos\beta) = \sum_{s=-\infty}^{\infty} i^s J_s(u)\exp(is\beta) \tag{6.2.18}$$

(see, for example, Gradshteyn and Ryzhik 1980; entry 8.511-4), where $J_s(u)$ is a Bessel function of order s. If we substitute this relation into eqn(6.2.17) and perform the integration over θ we obtain

$$\begin{aligned} C_j(\omega_1,\omega_2) &= \sum_{p=-\infty}^{\infty}\sum_{q=-\infty}^{\infty} i^{p+q} J_p(2n_j\omega_1)J_q(2n_j\omega_2) \\ &\times \frac{1}{2\pi}\int_{-\pi}^{\pi} \exp[i(p+2q)\theta]d\theta && (6.2.19) \\ &= \sum_{p=-\infty}^{\infty}\sum_{q=-\infty}^{\infty} i^{p+q} J_p(2n_j\omega_1)J_q(2n_j\omega_2)\delta_{p,-2q} \\ &= \sum_{p=-\infty}^{\infty} i^{3p} J_{2p}(2n_j\omega_1)J_p(2n_j\omega_2) && (6.2.20) \\ &\equiv R_j(\omega_1,\omega_2) + iI_j(\omega_1,\omega_2), && (6.2.21) \end{aligned}$$

where

$$R_j(\omega_1,\omega_2) = J_0(2n_j\omega_1)J_0(2n_j\omega_2) + 2\sum_{m=1}^{\infty}(-1)^m J_{4m}(2n_j\omega_1)J_{2m}(2n_j\omega_2) \tag{6.2.22}$$

and

$$I_j(\omega_1,\omega_2) = 2\sum_{m=0}^{\infty}(-1)^{m+1} J_{4m+2}(2n_j\omega_1)J_{2m+1}(2n_j\omega_2). \tag{6.2.23}$$

In deriving eqn(6.2.20) we have used the property $J_{-2q}(x) = J_{2q}(x)$.

The characteristic function corresponding to the required pdf is now obtained by substituting eqns (6.2.21) – (6.2.23) into eqn(6.2.14). The Fourier coefficients to be used with eqn(6.2.6) are obtained on making the replacements $\omega_1 \to \pi\alpha s$, $\omega_2 \to \pi\alpha t$ and introducing the abbreviations

$$R_{jst} = R_j(\pi\alpha s, \pi\alpha t) \text{ and } I_{jst} = I_j(\pi\alpha s, \pi\alpha t). \tag{6.2.24}$$

The Fourier coefficient becomes

$$C_{st} = \prod_{j=1}^{N/2}(R_{jst} + iI_{jst}) \equiv R_{st} + iI_{st}. \tag{6.2.25}$$

The Fourier series in eqn(6.2.6) can be simplified by exploiting the symmetry of the coefficients with respect to parity changes of the summation indices. Thus, I_{st} is a sum of products, each consisting of an odd number of the I_{jst} and some number of R_{jst}. Similarly, R_{st} is a sum of all products, each consisting of an even number of the I_{jst} and some number of R_{jst}. Consequently, one can determine the symmetry properties of R_{st} and I_{st} in terms of those of R_{jst} and I_{jst}. The explicit expressions (6.2.22) and (6.2.23) allow us to write

$$\begin{aligned} R_{j\bar{s}t} &= R_{js\bar{t}} = R_{j\bar{s}\bar{t}} = R_{jst} \\ I_{j\bar{s}t} &= I_{jst} \text{ and } I_{js\bar{t}} = -I_{jst}, \end{aligned} \tag{6.2.26}$$

which leads to

$$\begin{aligned} R_{\bar{s}t} &= R_{s\bar{t}} = R_{\bar{s}\bar{t}} = R_{st} \\ I_{\bar{s}t} &= I_{st} \text{ and } I_{s\bar{t}} = -I_{st}. \end{aligned} \tag{6.2.27}$$

A systematic introduction of these relations into the Fourier series allows us to simplify and reduce these series to the form

$$p(E_{\mathbf{h}}, E_{2\mathbf{h}}) = \left(\frac{\alpha^2}{2}\right)[G(E_{\mathbf{h}}, E_{2\mathbf{h}}) + U(E_{\mathbf{h}}, E_{2\mathbf{h}})], \tag{6.2.28}$$

where

$$\begin{aligned} G(E_{\mathbf{h}}, E_{2\mathbf{h}}) = 1 &+ 2\sum_{s=1}^{\infty}[R_{s0}\cos(\pi\alpha s E_{\mathbf{h}}) + R_{0s}\cos(\pi\alpha s E_{2\mathbf{h}})] \\ &+ 4\sum_{s=1}^{\infty}\sum_{t=1}^{\infty} R_{st}\cos(\pi\alpha s E_{\mathbf{h}})\cos(\pi\alpha t E_{2\mathbf{h}}) \end{aligned} \tag{6.2.29}$$

and

$$U(E_{\mathbf{h}}, E_{2\mathbf{h}}) = 4\sum_{s=1}^{\infty}\sum_{t=1}^{\infty} I_{st}\cos(\pi\alpha s E_{\mathbf{h}})\sin(\pi\alpha t E_{2\mathbf{h}}). \tag{6.2.30}$$

Clearly, the G component of the pdf is unaffected by the change of the sign of $E_{2\mathbf{h}}$, while if $E_{2\mathbf{h}}$ changes its sign so does the U component of the joint pdf (6.2.28). The ratio of the probabilities that $E_{2\mathbf{h}}$ is negative and positive is therefore given by

$$\frac{p_-}{p_+} = \frac{G(E_{\mathbf{h}}, E_{2\mathbf{h}}) - U(E_{\mathbf{h}}, |E_{2\mathbf{h}}|)}{G(E_{\mathbf{h}}, E_{2\mathbf{h}}) + U(E_{\mathbf{h}}, |E_{2\mathbf{h}}|}, \tag{6.2.31}$$

where $p_- = 1 - p_+$. Since

$$p_+ = \frac{p_+}{p_+ + p_-} = \left(1 + \frac{p_-}{p_+}\right)^{-1} \tag{6.2.32}$$

(Bertaut 1955), the probability that $E_{2\mathbf{h}}$ is positive, conditioned on the knowledge of $|E_{\mathbf{h}}|$ and $|E_{2\mathbf{h}}|$, is given by

$$p_+(2\mathbf{h}|\mathbf{h}) = \frac{1}{2}\left(1 + \frac{U(E_{\mathbf{h}}, |E_{2\mathbf{h}}|)}{G(E_{\mathbf{h}}, E_{2\mathbf{h}})}\right). \tag{6.2.33}$$

The functions G and U are defined by eqns (6.2.29) and (6.2.30) and depend in eqn(6.2.33) on the magnitudes of E alone.

6.2.2 *Reduction of the exact result to the hyperbolic tangent formula*

The conditional probability for the Σ_1 relationship given by eqn(6.2.33) is exact. We can, however, recover the corresponding central limit theorem approximation, first derived by Vand and Pepinsky (1953) and Cochran and Woolfson (1955), directly from the characteristic function.

We have shown at the beginning of Chapter 3 that the central limit theorem and its extensions can be derived from the properties of the characteristic function in the limit $\omega \to 0$. In practice, this is equivalent to retaining the lowest-order terms in an expansion of the characteristic function, and exponentiating them. Let us therefore approximate the characteristic function for the joint pdf (6.2.6) by considering the lowest-order terms in $R_j(\omega_1, \omega_2)$ and $I_j(\omega_1, \omega_2)$:

$$\begin{aligned} R_j(\omega_1, \omega_2) &\sim J_0(2n_j\omega_1)J_0(2n_j\omega_2) \\ &\sim (1 - n_j^2\omega_1^2)(1 - n_j^2\omega_2^2) \\ &\sim 1 - n_j^2(\omega_1^2 + \omega_2^2) \end{aligned} \tag{6.2.34}$$

and

$$\begin{aligned} I_j(\omega_1, \omega_2) &\sim J_2(2n_j\omega_1)J_1(2n_j\omega_2) \\ &\sim -n_j^3\omega_1^2\omega_2. \end{aligned} \tag{6.2.35}$$

The remaining Bessel functions give rise to higher-order corrections and are therefore omitted. If we use these results we can write for the lowest-order terms in the atomic contribution to the characteristic function

$$\begin{aligned} C_j(\omega_1, \omega_2) &\sim 1 - n_j^2(\omega_1^2 + \omega_2^2) - in_j^3\omega_1^2\omega_2 \\ &\sim \exp[-n_j^2(\omega_1^2 + \omega_2^2) - in_j^3\omega_1^2\omega_2], \end{aligned} \tag{6.2.36}$$

where the approximation $1 - z \sim \exp(-z)$, useful for very small z, has been employed. If we substitute eqn(6.2.36) into eqn(6.2.14), we find an expansion for $C(\omega_1, \omega_2)$ which contains terms up to the third order in the ωs:

$$
\begin{aligned}
C(\omega_1, \omega_2) &\simeq \prod_{j=1}^{N/2} \exp[-n_j^2(\omega_1^2+\omega_2^2) - i n_j^3 \omega_1^2 \omega_2] \\
&= \exp\left[-\left(\sum_{j=1}^{N/2} n_j^2\right)(\omega_1^2+\omega_2^2) - i\left(\sum_{j=1}^{N/2} n_j^3\right)\omega_1^2\omega_2 \right] \\
&= \exp\left[-\frac{1}{2}(\omega_1^2+\omega_2^2) - i\frac{\sigma_3}{2\sigma_2^{3/2}}\omega_1^2\omega_2 \right], \qquad (6.2.37)
\end{aligned}
$$

where we have made use of the relations: $\sum_{j=1}^{N} n_j^2 = 1$, $\sum_{j=1}^{N} n_j^3 = \sigma_3/\sigma_2^{3/2}$ and $\sigma_m = \sum_{j=1}^{N} f_j^m$. Since the the third-order terms are of lower order of magnitude than the quadratic ones we may reexpand the exponential containing them. In this way we find

$$
C(\omega_1, \omega_2) \sim \exp\left[-\frac{1}{2}(\omega_1^2+\omega_2^2) \right] \left[1 - i\frac{\sigma_3}{2\sigma_2^{3/2}}\omega_1^2\omega_2 \right]. \qquad (6.2.38)
$$

We can now obtain the asymptotic joint pdf by inverting the characteristic function in eqn(6.2.38)

$$
\begin{aligned}
p(E_{\mathbf{h}}, E_{2\mathbf{h}}) &\simeq \frac{1}{4\pi^2}\int_{-\infty}^{\infty}\int_{-\infty}^{\infty} C(\omega_1,\omega_2)\exp[-i(\omega_1 E_{\mathbf{h}} + \omega_2 E_{2\mathbf{h}})]d\omega_1 d\omega_2 \\
&\simeq \frac{1}{2\pi}\exp\left[-\frac{1}{2}(E_{\mathbf{h}}^2 + E_{2\mathbf{h}}^2) \right]\left[1 + \frac{\sigma_3}{2\sigma_2^{3/2}} E_{2\mathbf{h}}(E_{\mathbf{h}}^2 - 1) \right].
\end{aligned}
$$
(6.2.39)

If the last factor, in square brackets, is again approximated by an exponential, we obtain

$$
p(E_{\mathbf{h}}, E_{2\mathbf{h}}) \sim \frac{1}{2\pi}\exp\left[-\frac{1}{2}(E_{\mathbf{h}}^2 + E_{2\mathbf{h}}^2) + \frac{\sigma_3}{2\sigma_2^{3/2}} E_{2\mathbf{h}}(E_{\mathbf{h}}^2 - 1) \right]. \qquad (6.2.40)
$$

The classical hyperbolic tangent formula, shown in eqn(6.2.2), can now be derived from eqn(6.2.40) in a manner analogous to that leading to eqn(6.2.33). The ratio of the probabilities that the sign of $E_{2\mathbf{h}}$ is negative and positive is given by

$$
\frac{p_-}{p_+} \simeq \exp(-x) \qquad (6.2.41)
$$

where $x = \sigma_3 \sigma_2^{-3/2}|E_{2\mathbf{h}}|(E_{\mathbf{h}}^2 - 1)$. If we now make use of Bertaut's identity [eqn(6.2.32)], we obtain

$$p_+ = \left(1 + \frac{p_-}{p_+}\right)^{-1} \simeq [1 + \exp(-x)]^{-1}$$

$$= \left(\frac{\exp(x/2)}{\exp(x/2) + \exp(-x/2)}\right) \quad (6.2.42)$$

$$= \tanh(x/2) + 1 - p_+ \quad (6.2.43)$$

and solving for p_+ we have

$$p_+ \simeq \frac{1}{2} + \frac{1}{2}\tanh(x/2) = \frac{1}{2} + \frac{1}{2}\tanh\left[\frac{\sigma_3}{2\sigma_2^{3/2}}|E_{2\mathbf{h}}|(E_{\mathbf{h}}^2 - 1)\right], \quad (6.2.44)$$

which is the required result.

The linear approximation to the hyperbolic tangent formula

$$p_+(2\mathbf{h}|\mathbf{h}) \sim \frac{1}{2} + \frac{\sigma_3}{4\sigma_2^{3/2}}|E_{2\mathbf{h}}|(E_{\mathbf{h}}^2 - 1) \quad (6.2.45)$$

(Hauptman and Karle 1953*b*) can be derived directly from eqn(6.2.44) by expanding it in a MacLaurin series (Giacovazzo 1980).

6.2.3 *Illustrative examples*

Our comparison of the exact and approximate expressions for the conditional probability for the Σ_1 relationship deals with three main types of effects:

1. The effect of the magnitudes of $E_{\mathbf{h}}$ and $E_{2\mathbf{h}}$ on the computed probabilities is of obvious interest. As can be foreseen from the simplest Harker and Kasper (1948) inequality [see eqn(6.2.1)], when $|E_{\mathbf{h}}|$ and $|E_{2\mathbf{h}}|$ are very large the probability that $E_{2\mathbf{h}}$ is positive is very close to 1. It is therefore of particular interest to compare the predictions of the deterministic inequality with the exact conditional probability for the positive sign of $E_{2\mathbf{h}}$.
2. We are also interested in the effect of the number of atoms in the asymmetric unit on the discrepancy between the exact and approximate values of $p_+(2\mathbf{h}|\mathbf{h})$. Of course, the concept of 'number of atoms' in the present context is meaningful only if no outstandingly heavy atoms are present. Thus, when the number of equal atoms is large, we expect the central limit theorem to work well, which would lead to a good agreement between the approximate and exact probabilities for the positive sign of $E_{2\mathbf{h}}$. This agreement clearly depends on the magnitudes of the Es involved.
3. Another important effect is the degree of atomic heterogeneity. The composition of the asymmetric unit is taken as C_pX_q and the

composition-dependent quantities are expressed in terms of p, q and ρ, where ρ is the ratio of the atomic numbers Z_C and Z_X.

This notation allows us to write

$$n_C = \frac{f_C}{[2(pf_C^2 + qf_X^2)]^{1/2}} \simeq \frac{Z_C}{[2(pZ_C^2 + qZ_X^2)]^{1/2}} = \frac{1}{[2(p + q\rho^2)]^{1/2}},$$

$$n_X = \frac{f_X}{[2(pf_C^2 + qf_X^2)]^{1/2}} \simeq \frac{Z_X}{[2(pZ_C^2 + qZ_X^2)]^{1/2}} = \frac{\rho}{[2(p + q\rho^2)]^{1/2}},$$

$$\alpha = \frac{1}{E_{max}} = [2(pn_C + qn_X)]^{-1}$$

and

$$\frac{\sigma_3}{2\sigma_2^{3/2}} = \frac{p + q\rho^3}{[2(p + q\rho^2)]^{3/2}}.$$

The drawings shown in Figure 6.1 illustrate the effects of the magnitude of the $|E|$ values. Figure 6.1*a* shows the probability for the positive sign of $E_{2\mathbf{h}}$ as a function of $|E_{\mathbf{h}}|$ for a fixed atomic composition of the asymmetric unit, $C_{30}X_2$ with $Z_X/Z_C = 8$, as computed from the exact equation (6.2.33) (solid line) and from the approximate classical hyperbolic tangent equation (dashed line). The magnitude $|E_{2\mathbf{h}}|$ is taken as 1.75. It is evident that the approximate probability significantly underestimates that computed from the exact expression in the interesting range $|E_{\mathbf{h}}| > 1.2$ and has a qualitatively different behaviour for values of $|E_{\mathbf{h}}|$ lower than 1.2. The discrepancy between these two representations increases with an increasing degree of atomic heterogeneity. On the other hand, in the equal-atom case, the above two representations of the probability for the positive sign of $E_{2\mathbf{h}}$ give roughly the same results provided the number of atoms in the asymmetric unit is not too small. Figure 6.1*b* shows these representations for the same values of $|E_{\mathbf{h}}|$ and $|E_{2\mathbf{h}}|$ as in Fig. 6.1*a* and an equal-atom (C_{32}) asymmetric unit. The agreement between the two representations is excellent at low and intermediate values of E. A minor discrepancy appears at higher values of $|E_{\mathbf{h}}|$.

The effect of the number of atoms in the asymmetric unit of $P\bar{1}$ on the probability of a positive sign of the semi-invariant $E_{2\mathbf{h}}$ is illustrated in Fig. 6.2. In that figure the $|E|$s are fixed at the moderately large values $|E_{\mathbf{h}}| = |E_{2\mathbf{h}}| = 2.25$ and the probabilities are computed for the composition C_nX_2 as a function of the number of carbons, n. Figure 6.2*a* shows the result for the equal-atom case, while the composition represented in Fig. 6.2b is C_nKr_2. The exact probability is denoted by a solid line, and the approximation by a dashed one. Figure 6.2a displays a significant discrepancy between the exact and approximate probabilities, caused by a small number of atoms, in the equal-atom case, along with a tendency at large

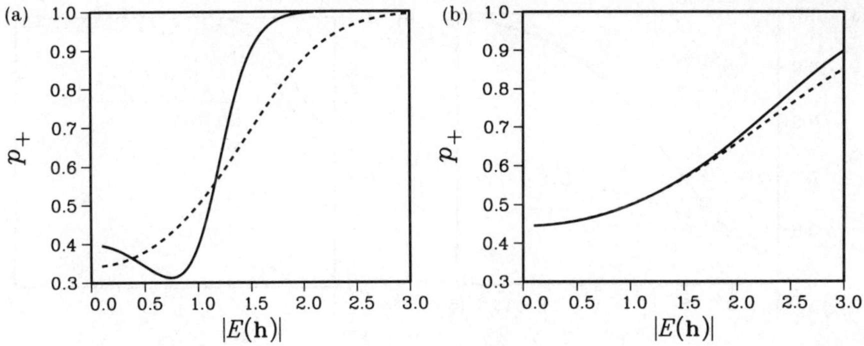

FIG. 6.1. *Effect of the magnitude of* $|E_{\mathbf{h}}|$ *on* $p_+(2\mathbf{h}|\mathbf{h})$. $|E_{2\mathbf{h}}|$ is fixed at 1.75. Asymmetric unit contents: (a) $C_{30}Cd_2$, (b) C_{32}.

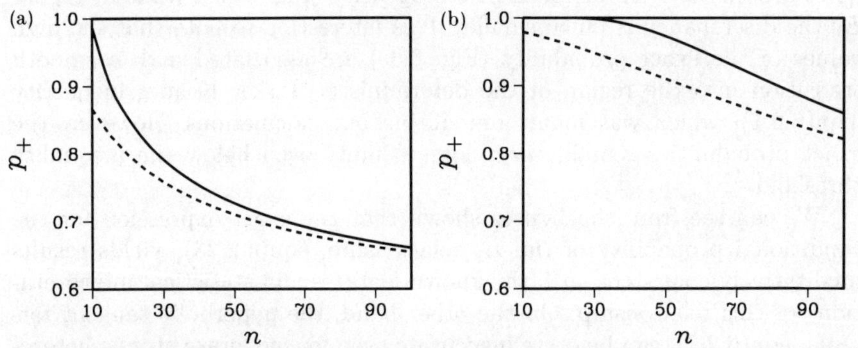

FIG. 6.2. *Effect of the number of (light) atoms on* $p_+(2\mathbf{h}|\mathbf{h})$. Both $|E_{\mathbf{h}}|$ and $|E_{2\mathbf{h}}|$ are fixed at 2.25. Asymmetric unit: (a) C_{n+2}, (b) C_nKr_2.

$|E|$ values to bring both representations into agreement as the number of the equal atoms increases. The presence of atomic heterogeneity affects the discrepancy between the two representations as shown in Fig. 6.2b. It appears that in the case represented in this figure the range of n over which the discrepancy persists is more extended than in the equal-atom case.

The effects of varying atomic heterogeneity, the other parameters being kept constant, are illustrated by Fig. 6.3. In this figure the test asymmetric unit of space group $P\bar{1}$ has the composition $C_{30}X_2$, with $\rho = Z_X/Z_C$ varying from 1 to 15. The $|E|$ values used in Figs 6.3*a* and 6.3*b* are $|E_{\mathbf{h}}| = |E_{2\mathbf{h}}| = 1.50$ and 2.25 respectively. Again, the exact and approximate representations are denoted by solid and dashed lines respectively. At

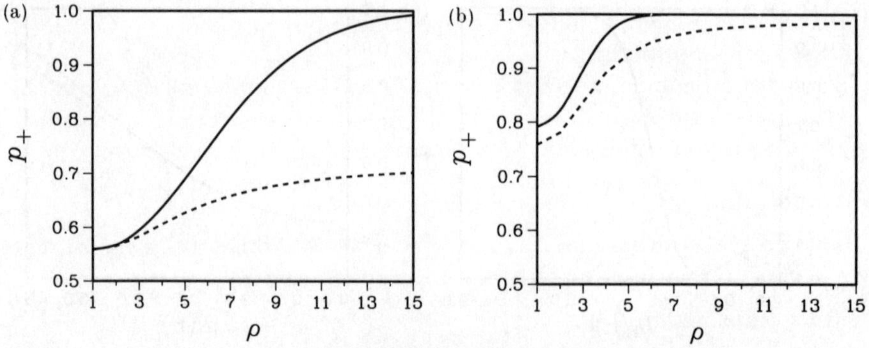

FIG. 6.3. *Effect of atomic heterogeneity on* $p_+(2\mathbf{h}|\mathbf{h})$ *for selected E values.* (a) $|E_\mathbf{h}| = |E_{2\mathbf{h}}| = 1.50$. (b) $|E_\mathbf{h}| = |E_{2\mathbf{h}}| = 2.25$.

low values of $|E|$ the discrepancy between the two representations increases quite drastically with increasing heterogeneity (Fig. 6.3*a*), while at higher Es the discrepancy is much smaller. It is interesting to note that the high values of the exact probability (Fig. 6.3*b*) are associated with a smooth transition into the region of the deterministic Harker–Kasper inequality [eqn(6.2.1)] which was monitored during our calculations. However, the exact probabilities remain very close to unity even below the inequality threshold.

We can see from the figures shown that the exact expression for the conditional probability of the Σ_1 relationship, eqn(6.2.33), yields results qualitatively consistent with the known features and statistical interpretations of this relationship. On the other hand, the hyperbolic tangent formula [eqn(6.2.2)] can be quite inaccurate even for moderate atomic heterogeneities and, more importantly, for the equal-atom case when the number of atoms in the asymmetric unit is small.

The preceding illustrations, and calculations for other choices of parameters, indicate that the discrepancies between the exact and approximate expressions for the conditional probability for the positive sign of $E_{2\mathbf{h}}$ decrease as the product $K = |E_{2\mathbf{h}}|[|E_\mathbf{h}|^2 - 1]$ increases. The region of greatest discrepancy, for a given K, seems to depend on the value of K itself. Thus, the discrepancy is greatest at large atomic heterogeneities for low values of K (Fig. 6.3a), and the range of maximum discrepancy moves to lower heterogeneities as K increases.

A result of immediate practical significance is the fact that the universally computed (more often computed than used) hyperbolic tangent formula (6.2.2) often gives what are regarded as unacceptably low indications, in situations in which the values of the exact probability, as given by eqn (6.2.33), would most probably be considered as acceptable. It therefore appears that the use of the exact eqn(6.2.33) in estimating the reliability of Σ_1 relationships should very significantly promote the determination of the signs of structure semi-invariants in the space group $P\bar{1}$, whenever significant atomic heterogeneity is present or the number of atoms in the asymmetric unit is small. It should also be pointed out that the computation of the exact conditional probability for Σ_1 is fairly easy. It appears to be feasible to make use of these results in many routine applications.

6.3 Exact conditional probability for the positive sign of the triple product $E_{\mathbf{h}}E_{\mathbf{k}}E_{\mathbf{h+k}}$ in space group $P\bar{1}$

6.3.1 *Introduction*

The previous section contains a detailed outline of the mathematical technique that can be used to construct joint pdfs of several structure factors, and a simple but non-trivial application of the formalism. This application, however, is confined to estimating the signs of reflections with all their indices being even. Only the probability for a positive sign of such a reflection is predictable.

The present section deals with a more general relationship, also belonging to the foundations of direct methods, on the basis of which the sign of any reflection can be estimated or predicted. The sign relationship to be discussed dates back to the work of Sayre (1952) who showed that for a centrosymmetric structure the product $F_{\mathbf{h}}F_{\mathbf{k}}F_{\mathbf{h+k}}$ tends to be positive. If we denote the sign of the structure factor of reflection $\mathbf{h}$ by $s_{\mathbf{h}}$, this relationship takes the form

$$s_{\mathbf{h}}s_{\mathbf{k}}s_{\mathbf{h+k}} \text{ probably equals } +1. \tag{6.3.46}$$

As in the case of Σ_1, the probability for the sign of the triple product becomes certainty when the magnitudes of the structure factors are large enough to satisfy the relevant Harker–Kasper inequality

$$(|U_{\mathbf{h}}| + |U_{\mathbf{k}}|)^2 \leq \{1 + s_{\mathbf{h}}s_{\mathbf{k}}s_{\mathbf{h+k}}|U_{\mathbf{h+k}}|\}\{1 + s_{\mathbf{h}}s_{\mathbf{k}}s_{\mathbf{h-k}}|U_{\mathbf{h-k}}|\} \tag{6.3.47}$$

(e.g. Woolfson 1970). More often than not, however, this is not the case and the probability for the sign of the product in eqn(6.3.46) must be found, conditioned on the magnitudes of the three structure factors involved. The simplest probabilistic treatment, valid for structures containing a large number of not too dissimilar atoms, is based on the central limit theorem (Woolfson 1954; Cochran and Woolfson 1955). The resulting approximate probability, associated with eqn(6.3.46) and represented in terms of normalized structure factors, is then given by

$$p_+(\mathbf{h},\mathbf{k},\mathbf{h}+\mathbf{k}) \simeq \frac{1}{2} + \frac{1}{2}\tanh\left(\frac{\sigma_3}{\sigma_2^{3/2}}|E_\mathbf{h}E_\mathbf{k}E_{\mathbf{h}+\mathbf{k}}|\right). \tag{6.3.48}$$

In the following subsection we derive an exact expression for this probability, and compare its performance with that of eqn(6.3.48). This derivation was published by Shmueli and Weiss (1986). We follow that reference in our present treatment.

6.3.2 *Derivation*

Our derivation is based on and is formally very similar to that for the Σ_1 relation presented in subsection 6.2.1. We first find the characteristic function for the joint pdf of the normalized structure factors, construct the corresponding Fourier series, and finally recover the conditional probability for the positive sign of the product of these structure factors. The derivation is carried out in detail for space group $P\bar{1}$.

The characteristic function for the joint pdf of the normalized structure factors $E_\mathbf{h}$, $E_\mathbf{k}$ and $E_{\mathbf{h}+\mathbf{k}}$, for the space group $P\bar{1}$, is given by

$$C(\boldsymbol{\omega}) = \langle\exp[i(\omega_1 E_\mathbf{h} + \omega_2 E_\mathbf{k} + \omega_3 E_{\mathbf{h}+\mathbf{k}}])\rangle \tag{6.3.49}$$

$$= \left\langle \exp\left\{2i\sum_{j=1}^{N/2} n_j[\omega_1\cos\theta_j + \omega_2\cos\phi_j + \omega_3\cos(\theta_j+\phi_j)]\right\}\right\rangle \tag{6.3.50}$$

$$\equiv \prod_{j=1}^{N/2} C_j(\boldsymbol{\omega}), \tag{6.3.51}$$

where N is the number of atoms in the unit cell, n_j is the normalized scattering factor of the jth atom, $\theta_j = 2\pi\mathbf{h}^T\mathbf{r}_j$, $\phi_j = 2\pi\mathbf{k}^T\mathbf{r}_j$ and

$$C_j(\boldsymbol{\omega}) = \frac{1}{(2\pi)^2}\int_{-\pi}^{\pi}\int_{-\pi}^{\pi}\exp\{2in_j[\omega_1\cos\theta + \omega_2\cos\phi + \omega_3\cos(\theta+\phi)]\}d\theta d\phi \tag{6.3.52}$$

is the contribution of the jth atom to the characteristic function. The integral in eqn(6.3.52) can be converted to series form with the aid of the identity given in eqn(6.2.18). We then obtain

$$C_j(\boldsymbol{\omega}) = \sum_p\sum_q\sum_r i^{p+q+r}J_p(2n_j\omega_1)J_q(2n_j\omega_2)J_r(2n_j\omega_3)$$
$$\times\left\{\frac{1}{(2\pi)^2}\int_{-\pi}^{\pi}\int_{-\pi}^{\pi}\exp\{i[(p+r)\theta + (q+r)\phi]\}d\theta d\phi\right\} \tag{6.3.53}$$
$$= \sum_p\sum_q\sum_r i^{p+q+r}J_p(2n_j\omega_1)J_q(2n_j\omega_2)J_r(2n_j\omega_3)\delta_{pr}\delta_{qr}$$

$$\tag{6.3.54}$$

$$= \sum_{p=-\infty}^{\infty} i^{3p} \prod_{k=1}^{3} J_p(2n_j\omega_k) \tag{6.3.55}$$

$$\equiv R_j(\boldsymbol{\omega}) + iI_j(\boldsymbol{\omega}) \tag{6.3.56}$$

where

$$R_j(\boldsymbol{\omega}) = \prod_{k=1}^{3} J_0(2n_j\omega_k) + 2\sum_{p=1}^{\infty}(-1)^p \prod_{k=1}^{3} J_{2p}(2n_j\omega_k) \tag{6.3.57}$$

and

$$I_j(\boldsymbol{\omega}) = 2\sum_{p=0}^{\infty}(-1)^{p+1} \prod_{k=1}^{3} J_{2p+1}(2n_j\omega_k). \tag{6.3.58}$$

The characteristic function (6.3.49) can now be written as

$$C(\boldsymbol{\omega}) = \prod_{j=1}^{N/2}[R_j(\boldsymbol{\omega}) + iI_j(\boldsymbol{\omega})] = R(\boldsymbol{\omega}) + iI(\boldsymbol{\omega}). \tag{6.3.59}$$

As in the previous section, the Fourier coefficients for the required joint pdf are obtained from the characteristic function by replacing the transform variables ω_k by quantities related to the summation indices. In our case, we put

$$\omega_1 \to \pi\alpha r,\ \omega_2 \to \pi\alpha s,\ \omega_3 \to \pi\alpha t, \tag{6.3.60}$$

where r, s, t are the indices of the triple Fourier series for the required pdf, and

$$\frac{1}{\alpha} = E_{max} = \sum_{j=1}^{N} n_j \tag{6.3.61}$$

is the maximum magnitude of E. If we write

$$R(\omega_1, \omega_2, \omega_3) = R(\pi\alpha r, \pi\alpha s, \pi\alpha t) \equiv R_{rst} \tag{6.3.62}$$

and similarly for I_{rst}, we can write the joint pdf as

$$p(E_1, E_2, E_3) = \frac{\alpha^3}{8}\sum_r\sum_s\sum_t(R_{rst} + iI_{rst})\exp[-i(rE_1 + sE_2 + tE_3)] \tag{6.3.63}$$

where $E_1 = \pi\alpha E_{\mathbf{h}}$, $E_2 = \pi\alpha E_{\mathbf{k}}$ and $E_3 = \pi\alpha E_{\mathbf{h+k}}$. Of course, only the real part of eqn(6.3.63) is relevant.

We can reduce the Fourier series in eqn(6.3.63) to a sum over the non-negative octant of the *rst* index space by making use of the symmetry properties of the trigonometric functions, as well as those of R_{rst} and I_{rst}. If we observe that $J_{-n}(x) = (-1)^n J_n(x)$, $J_n(-x) = (-1)^n J_n(x)$ and take account of the detailed form of eqns (6.3.57) and (6.3.58), we conclude that

1. R_{rst} is symmetric in all three indices (i.e. invariant to their sign changes), and
2. I_{rst} changes its sign when one or three indices change sign(s), and retains its sign when the signs of any two indices are changed.

The consequent simplification of eqn(6.3.63) follows in much the same way as the simplification of the electron density for given symmetries of the reciprocal lattice (*e.g. International Tables* 1965). If we draw an analogy between the rst and hkl Fourier spaces, R_{rst} and I_{rst} can be said to possess symmetries isomorphic to those of mmm and 222 respectively.

On making use of these symmetry considerations, we obtain

$$p(E_1, E_2, E_3) = \frac{\alpha^3}{8}(p_{even} + p_{odd}), \tag{6.3.64}$$

where

$$\begin{aligned} p_{even} &= 1 + 2\sum_{r=1}^{\infty} R_{r00}(C_{r1} + C_{r2} + C_{r3}) \\ &+ 4\sum_{r=1}^{\infty}\sum_{s=1}^{\infty} R_{rs0}(C_{r1}C_{s2} + C_{r2}C_{s3} + C_{r3}C_{s1}) \\ &+ 8\sum_{r=1}^{\infty}\sum_{s=1}^{\infty}\sum_{t=1}^{\infty} R_{rst}C_{r1}C_{s2}C_{t3} \end{aligned} \tag{6.3.65}$$

with $C_{wk} = \cos(wE_k), w = r, s, t$, and

$$p_{odd} = -8\sum_{r=1}^{\infty}\sum_{s=1}^{\infty}\sum_{t=1}^{\infty} I_{rst}S_{r1}S_{s2}S_{t3} \tag{6.3.66}$$

with $S_{wk} = \sin(wE_k), w = r, s, t$. We see that the p_{even} component of the joint pdf is invariant under the change of sign of the triple product, while the sign combinations '−++', '+−+', '++−' and '− − −', which lead to negative triple products, reverse the sign of p_{odd} in eqn(6.3.66). The ratio of the probabilities for negative and positive triple products is therefore given by

$$\frac{p_-(\mathbf{h}, \mathbf{k}, \mathbf{h}+\mathbf{k})}{p_+(\mathbf{h}, \mathbf{k}, \mathbf{h}+\mathbf{k})} = \frac{p_{even}(E_\mathbf{h}, E_\mathbf{k}, E_{\mathbf{h}+\mathbf{k}}) - p_{odd}(|E_\mathbf{h}|, |E_\mathbf{k}|, |E_{\mathbf{h}+\mathbf{k}}|)}{p_{even}(E_\mathbf{h}, E_\mathbf{k}, E_{\mathbf{h}+\mathbf{k}}) + p_{odd}(|E_\mathbf{h}|, |E_\mathbf{k}|, |E_{\mathbf{h}+\mathbf{k}}|)}. \tag{6.3.67}$$

On making use of Bertaut's identity in eqn(6.2.32) we find that the required probability that the triple product $E_\mathbf{h}E_\mathbf{k}E_{\mathbf{h}+\mathbf{k}}$ is positive is given by

$$p_+(\mathbf{h},\mathbf{k},\mathbf{h}+\mathbf{k}) = \frac{1}{2}\left[1 + \frac{p_{odd}(|E_\mathbf{h}|,|E_\mathbf{k}|,|E_{\mathbf{h}+\mathbf{k}}|)}{p_{even}(E_\mathbf{h},E_\mathbf{k},E_{\mathbf{h}+\mathbf{k}})}\right], \tag{6.3.68}$$

where p_{even} and p_{odd} are defined by eqns (6.3.65) and (6.3.66) respectively.

We wish to point out that, since for any non-zero integer n, $J_n(0) = 0$ and $J_0(0) = 1$, the special coefficients in eqn(6.3.65) are given by

$$R_{r00} = \prod_{j=1}^{N/2} J_0(2\pi\alpha r n_j) \tag{6.3.69}$$

and

$$R_{rs0} = \prod_{j=1}^{N/2} J_0 2\pi\alpha r n_j) J_0(2\pi\alpha s n_j) \tag{6.3.70}$$

since the remaining infinite series of Bessel functions in eqn(6.3.57) then vanishes identically.

Equations (6.3.64) and (6.3.67) are exact. It is possible to rederive from these results the classical hyperbolic tangent formula, eqn(6.3.48), by following a procedure similar to that outlined in subsection 6.2.2. As in that subsection, the derivation of eqn(6.3.48) is based on an approximation to the characteristic function by its lowest-order term and a recalculation of the approximate joint pdf by a Fourier inversion.

6.3.3 *Illustrative examples*

We now compare the results corresponding to eqn(6.3.68) with the hyperbolic tangent approximation, given by eqn(6.3.48), for several hypothetical $P\bar{1}$ structures. As in the previous section, which dealt with the Σ_1 relationship, we shall try to examine here the effects of (i) the number of atoms in the asymmetric unit, (ii) the magnitudes of the E values and (iii) atomic heterogeneity of the structure. All these effects are, of course, interrelated.

Figure 6.4 illustrates the effect of the paucity of atoms for some selected $|E|$ values. The assumed composition of the asymmetric unit of $P\bar{1}$ is C_{10}. In the figure $|E_\mathbf{h}|$ and $|E_\mathbf{k}|$ are kept fixed and $|E_{\mathbf{h}+\mathbf{k}}|$ ranges from 1 to 3. The often used threshold of acceptability, $p_+ = 0.95$, is denoted by a dashed-dotted horizontal line, extending across all the figures in this section. The probability computed from the exact eqn(6.3.68) is denoted by solid lines, while that given by the approximate hyperbolic tangent formula (6.3.48) is drawn with dashed lines.

Figure 6.4*a* shows that both probabilities decrease as the value of the triple product decreases, and the approximate probability significantly underestimates the exact one. In Figure 6.4*b* the exact probabilities are essentially equal to one throughout the range of $|E_{\mathbf{h}+\mathbf{k}}|$. This is to be expected on the basis of the related Harker–Kasper inequality (Harker and Kasper

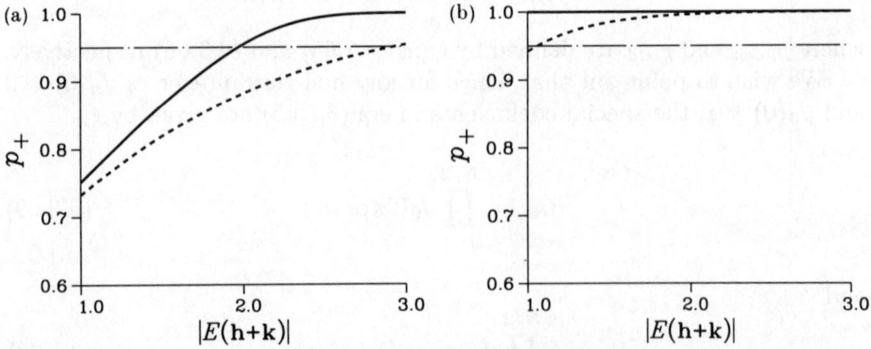

FIG. 6.4. *Effect of a small number of atoms on the probability p_+ for the positive sign of the triple product.* (a) $|E_\mathbf{h}| = |E_\mathbf{k}| = 1.50$. (b) $|E_\mathbf{h}| = |E_\mathbf{k}| = 2.50$.

1948). The hyperbolic tangent formula again underestimates the exact result at lower values of $|E_{\mathbf{h+k}}|$.

Figures 6.5*a* and 6.5*b* contain curves of the probability, as a function of $|E_{\mathbf{h+k}}|$, that the sign of the triple product is positive, the compositions of the asymmetric unit being C_{32} and $C_{32}Kr_2$ respectively. We assume that $|E_\mathbf{h}| = |E_\mathbf{k}| = 1.50$. For the equal-atom case (Fig. 6.5*a*) the exact and approximate probabilities are rather close each to the other, the discrepancy increasing slightly as the triple product increases. The replacement of two carbons by two kryptons ($Z = 36$) strongly enhances both the exact and the approximate probabilities, but also leads to large discrepancies at low values of the triple product. We finally illustrate the effect of a small number of very heavy atoms in a moderately large asymmetric unit, having the composition $C_{95}U_5$, for two selected choices of $|E_\mathbf{h}|$ and $|E_\mathbf{k}|$. Although the contribution of the uranium is dominant, the discrepancies between exact and approximate probabilities, at low values of the triple product, are significantly smaller than those obtained for an equal-atom structure with five atoms in the asymmetric unit. The known favourable effect of the heavy atoms on correct sign indications is consistent with the fact that probabilities computed from the exact expression in eqn(6.3.49) are higher than those obtained from the approximate one in eqn(6.3.48).

Summarizing, these and other numerical examples indicate that the approximate probabilities for the positive sign of the triple product based on the central limit theorem consistently underestimate the exact ones given by their Fourier representations. The discrepancies between these two probabilities increase as the atomic heterogeneity increases or the number of atoms in the asymmetric unit of space group $P\bar{1}$ decreases. Since the ex-

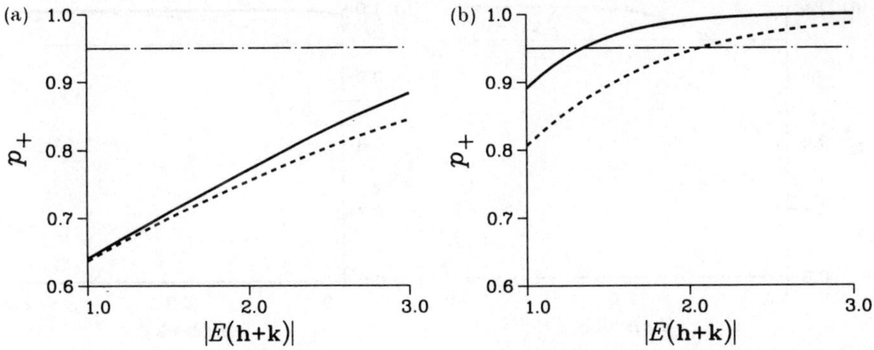

FIG. 6.5. *Effect of atomic heterogeneity on the probability for the positive sign of the triple product.* $|E_{\mathbf{h}}| = |E_{\mathbf{k}}| = 1.50$. Asymmetric unit contents: (a) C_{32},(b) $C_{30}Kr_2$.

act probability as given by eqn(6.3.68) is rather rapidly computed — and this can be speeded up by the use of the Fast Fourier Transform algorithm (Cooley and Tukey 1965) — an implementation of eqn(6.3.68) in crystallographic application software is a practical proposition which may enhance the reliability of sign determination during a construction of the starting set. No less important is the role of eqn(6.3.68) as a standard of comparison with various existing approximations, and a basis for construction of new and readily computable approximations based on this exact result. Such an approximation has not yet been constructed for eqn(6.3.68), but would be both useful and feasible. The next section, which deals with a much more complicated structure invariant, contains an example of such simple approximation to the exact result.

6.4 Exact conditional pdf of the three-phase invariant in space group $P1$

6.4.1 *N-phase invariants*

Direct methods of phase determination are based on the the use of linear combinations of phases which are independent of the choice of the origin of the unit cell, generally known as structure invariants. The important concept of structure invariance was first formulated rigorously by Hauptman and Karle (1953*b*). We wish to recall here, by way of introduction, how these invariants are derived and then proceed to a derivation of their probabilistic properties.

Consider the product of n normalized structure factors, for a crystal structure belonging to the simplest (non-centrosymmetric) space group $P1$. This product can be written as

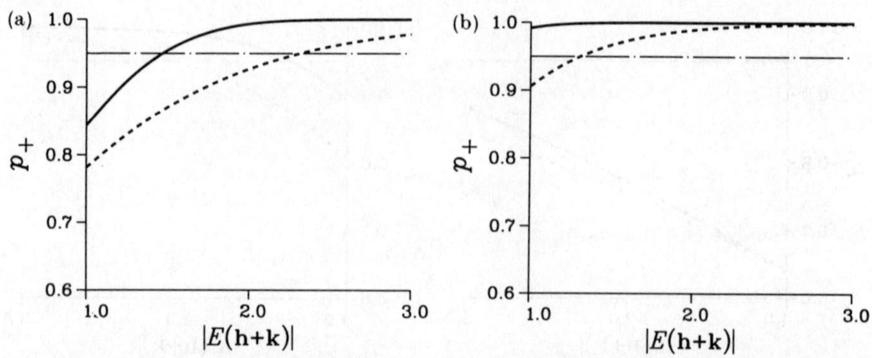

FIG. 6.6. *A small number of very heavy atoms in a moderately large asymmetric unit.* Asymmetric unit contents: $C_{95}U_5$. (a) $|E_{\mathbf{h}}| = |E_{\mathbf{k}}| = 1.50$. (b) $|E_{\mathbf{h}}| = |E_{\mathbf{k}}| = 2.00$.

$$\prod_{k=1}^{n} E(\mathbf{h}_k) = \prod_{k=1}^{n} \left\{ \sum_{j=1}^{N} n_j \exp(2\pi \mathbf{h}_k^T \mathbf{r}_j) \right\} \tag{6.4.71}$$

$$= \left[\prod_{k=1}^{n} |E(\mathbf{h}_k)| \right] \exp \left[i \sum_{k=1}^{n} \varphi(\mathbf{h}_k) \right], \tag{6.4.72}$$

where the amplitude-phase description of the structure factor, i.e.

$$E(\mathbf{h}_k) = |E(\mathbf{h}_k)| \exp[i\varphi(\mathbf{h}_k)]$$

has been used. Suppose now that the origin to which the position vectors $\mathbf{r}_j$, in the right-hand side of eqn(6.4.71), are referred has been shifted so that $\mathbf{r}_j$ is replaced by $\mathbf{r}_j - \mathbf{d}$, for each j. This shift of the origin must be accompanied by a shift of the phase of the structure factor $E(\mathbf{h}_k)$ by $-2\pi \mathbf{h}_k^T \mathbf{d}$. The phase of the product of the n structure factors now becomes

$$\sum_{k=1}^{n} \left[\varphi(\mathbf{h}_k) - 2\pi \mathbf{h}_k^T \mathbf{d} \right]. \tag{6.4.73}$$

The magnitudes of E, in eqn(6.4.72), are of course independent of the choice of the unit cell origin. Hence, possible values of the diffraction vectors $\mathbf{h}_k$ for which the phase of the product (6.4.72) is independent of the shift of the origin, for any $\mathbf{d}$, are those that satisfy the relationships

$$\sum_{k=1}^{n} \mathbf{h}_k = \mathbf{0} \tag{6.4.74}$$

or

$$\mathbf{h}_n = -\sum_{k=1}^{n-1} \mathbf{h}_k. \tag{6.4.75}$$

In what follows we discuss, neglecting effects of dispersion, the simplest non-trivial phase invariant, i.e. the three-phase invariant. It follows that the product

$$E(\mathbf{h})E(\mathbf{k})E(-\mathbf{h}-\mathbf{k}), \tag{6.4.76}$$

and therefore the phase sum

$$\Phi = \varphi(\mathbf{h}) + \varphi(\mathbf{k}) + \varphi(-\mathbf{h}-\mathbf{k}) \tag{6.4.77}$$

are independent of choice of the space-group origin, i.e. are examples of structure invariants. Incidentally, the triple product discussed in the last section is also a structure invariant and its phase can be either 0 or π. This follows simply from the fact that for centrosymmetric crystals $E(-\mathbf{h}) = E(\mathbf{h})$.

A conditional pdf of Φ, the *three-phase invariant*, was first derived using the central limit theorem by Cochran (1955) (see also section 3.5 above). Corrections to Cochran's result were subsequently derived by a number of investigators in terms of Gram–Charlier, Edgeworth or exponentiated series [e.g. Naya, Nitta and Oda (1965); Hauptman (1971); Karle (1972); Karle and Gilardi (1973); Giacovazzo (1974); Peschar and Schenk (1986)]. A rich source of references to such studies is the book of Giacovazzo (1980). Our treatment of this conditional pdf in this section is based on the work of Shmueli, Rabinovich and Weiss (1989*a*, 1989*b*) which, in turn, applies the method suggested by us for constructing Fourier representations of multivariate joint pdfs. We first outline our derivation of the conditional pdf of Φ for space group $P1$, indicate its reduction to the Cochran (1955) conditional pdf for $P1$ structures consisting of 'equal atoms and highly populated unit cells', and illustrate the results graphically. It will be seen that the exact formulation of the problem gives rise to rather time-consuming computations. We conclude by describing our attempts at coping with this difficulty by constructing rapidly computable approximations from the exact results.

6.4.2 *Derivation of the conditional joint pdf of Φ*

We first construct a joint probability density function (jpdf) of the real and imaginary parts of the three structure factors related to the three-phase invariant Φ. Let these structure factors be written as $E(\mathbf{h}) \equiv A_1 + iB_1$, $E(\mathbf{k}) \equiv A_2 + iB_2$ and $E(-\mathbf{h}-\mathbf{k}) \equiv A_3 + iB_3$. Since each of the As and Bs can differ from zero only in $(-1/\alpha, 1/\alpha)$, where $1/\alpha$ is the maximum value of $|E|$, we may express the required jpdf as the sixfold Fourier series

$$p(\mathbf{E}) = K \sum_{\mathbf{u}} C_{\mathbf{u}} \exp\left[-\pi i \alpha \sum_{k=1}^{3} (u_k A_k + v_k B_k)\right], \tag{6.4.78}$$

where

$$\mathbf{u}^T = (u_1, v_1, u_2, v_2, u_3, v_3) \tag{6.4.79}$$

is the vector of Fourier summation indices,

$$\mathbf{E}^T = (A_1, B_1, A_2, B_2, A_3, B_3) \tag{6.4.80}$$

is the vector of real and imaginary components of the structure factors, K is a normalization constant, and $C_{\mathbf{u}}$ is a Fourier coefficient. (The present notation for the Fourier summation indices differs from that in our articles on this subject, resulting in some simplification of the subscripts.) If we introduce the following abbreviations

$$E_1 \equiv |E(\mathbf{h})|,\ E_2 \equiv |E(\mathbf{k})|,\ E_3 \equiv |E(-\mathbf{h}-\mathbf{k})|$$

and

$$\varphi_1 \equiv \varphi(\mathbf{h}),\ \varphi_2 \equiv \varphi(\mathbf{k}),\ \varphi_3 \equiv \varphi(-\mathbf{h}-\mathbf{k}),$$

we can introduce the natural polar coordinates: $A_i = E_i \cos\varphi_i$ and $B_i = E_i \sin\varphi_i$ and rewrite eqn(6.4.78) as

$$p(\mathbf{E}) = K \sum_{\mathbf{u}} C_{\mathbf{u}} \exp\left[-\pi i \alpha \sum_{k=1}^{3} E_k (u_k \cos\varphi_k + v_k \sin\varphi_k)\right] \tag{6.4.81}$$

$$= K \sum_{\mathbf{u}} C_{\mathbf{u}} \exp\left[-i \sum_{k=1}^{3} \Omega_k \cos(\varphi_k - \Delta_k)\right], \tag{6.4.82}$$

where

$$\Omega_k = \pi \alpha E_k (u_k^2 + v_k^2)^{1/2} \tag{6.4.83}$$

and

$$\Delta_k = \tan^{-1}(v_k/u_k). \tag{6.4.84}$$

The jpdf in eqn(6.4.82) depends on the individual magnitudes and phases of the three structure factors. This jpdf can be made to depend on the three-phase invariant if we replace, for example, φ_3 with $\Phi - \varphi_1 - \varphi_2$. Assuming that we know nothing about the phases, except for those that fix the origin of the space group, we can assume that the φ_is are uniformly distributed in $(-\pi, \pi)$. The required pdf of Φ, conditioned on the knowledge of the magnitudes of the three structure factors, is obtained by integrating out the phases φ_1 and φ_2. The relevant integral can be evaluated numerically but it is convenient to represent its solution in terms of series of Bessel functions by making use of eqn(6.2.18). The function to be integrated over φ_1 and φ_2, after φ_3 was replaced as indicated above, is the exponential on the right-hand side of eqn(6.4.82). If we denote its integral by $Z(\Phi)$, we obtain

$$\begin{aligned} Z_{\mathbf{u}}(\Phi) &= \int_{-\pi}^{\pi} d\varphi_1 \int_{-\pi}^{\pi} d\varphi_2 \exp\left[-i\sum_{k=1}^{3} \Omega_k \cos(\varphi_k - \Delta_k)\right] \\ &= \sum_p \sum_q \sum_r i^{p+q+r} J_p(-\Omega_1) J_q(-\Omega_2) J_r(-\Omega_3) \\ &\times \exp[i(r\Phi - p\Delta_1 - q\Delta_2 - r\Delta_3)] \\ &\times \int_{-\pi}^{\pi}\int_{-\pi}^{\pi} \exp\{i[(p-r)\varphi_1 + (q-r)\varphi_2]\} d\varphi_1 d\varphi_2. \end{aligned} \tag{6.4.85}$$

The angular integral in the last line of eqn(6.4.85) vanishes unless both $p = r$ and $q = r$. Hence, if we observe that $J_n(-x) = (-1)^n J_n(x)$ and omit a factor of $4\pi^2$, the right-hand side of eqn(6.4.85) simplifies to

$$Z_{\mathbf{u}} = \sum_{p=-\infty}^{\infty} i^p \left[\prod_{k=1}^{3} J_p(\Omega_k)\right] \exp[ip(\Phi - \Delta)], \tag{6.4.86}$$

where $\Delta = \Delta_1 + \Delta_2 + \Delta_3$. The general form of the conditional pdf of the three-phase invariant is therefore given by

$$p(\Phi | E_1, E_2, E_3, \ldots) = K' \sum_{\mathbf{u}} C_{\mathbf{u}} Z_{\mathbf{u}}(\Phi), \tag{6.4.87}$$

where K' is a normalization constant, $Z_{\mathbf{u}}$ is defined by eqn(6.4.86), and the Fourier coefficient $C_{\mathbf{u}}$ is to be evaluated for the specific (crystallographic and/or non-crystallographic) symmetry and composition. An integration of eqn(6.4.87) with respect to Φ shows that the (real) normalization constant is given by

$$K' = \left\{2\pi\mathcal{R}\left[\sum_{\mathbf{u}} C_{\mathbf{u}} \left(\prod_{k=1}^{3} J_0(\Omega_k)\right)\right]\right\}^{-1}, \tag{6.4.88}$$

where $\mathcal{R}$ is the real part of the expression in brackets in eqn(6.4.88).

6.4.3 *Derivation of Fourier coefficients*

Since the Fourier coefficient $C_{\mathbf{u}}$ in eqn(6.4.87) is just the value of the characteristic function at a point depending in a certain way on the summation indices (e.g. Shmueli and Weiss 1985b), the general expression for this coefficient is given by

$$C_{\mathbf{u}} = \left\langle \exp\left[i \sum_{k=1}^{3} (\omega_{2k-1} A_k + \omega_{2k} B_k) \right] \right\rangle, \tag{6.4.89}$$

where $\omega_{2m-1} = \pi\alpha u_m$ and $\omega_{2m} = \pi\alpha v_m$. The expressions for A_k and B_k for the space groups $P1$ and $P2$ will now be used to derive the Fourier coefficients for these space groups.

SPACE GROUP $P1$

For the space group $P1$ we have

$$A_k = \sum_{j=1}^{N} \cos\theta_{jk} \text{ and } B_k = \sum_{j=1}^{N} \sin\theta_{jk}, \tag{6.4.90}$$

where $\theta_{jk} = 2\pi \mathbf{h}_k^T \mathbf{r}_j$, with $\mathbf{h}_1 = \mathbf{h}$, $\mathbf{h}_2 = \mathbf{k}$ and $\mathbf{h}_3 = -\mathbf{h}-\mathbf{k}$. If we now make use of the assumption that the atomic contributions to the structure factor are statistically independent and of eqn(6.4.90), we can rewrite eqn(6.4.89) as

$$C_{\mathbf{u}} = \prod_{j=1}^{N} \left\langle \exp\left[i \sum_{k=1}^{3} \Lambda_{jk} \cos(\theta_{jk} - \Delta_k) \right] \right\rangle, \tag{6.4.91}$$

where

$$\Lambda_{jk} = \pi\alpha n_j (u_k^2 + v_k^2)^{1/2} \tag{6.4.92}$$

and Δ_k is given by eqn(6.4.84). The average in eqn(6.4.91) can be calculated analogously to the integration of the phase factor shown in eqn(6.4.85). The exponential is a product of three terms of the form $\exp(ix\cos\beta)$ and each of these is replaced by a series of Bessel functions, such as that in eqn(6.2.18). In this way we find that function $C_{\mathbf{u}}$, for space group $P1$, is

$$C_{\mathbf{u}} = \prod_{j=1}^{N} C_{\mathbf{u},j}(\Delta), \tag{6.4.93}$$

where

$$C_{\mathbf{u},j}(\Delta) = \sum_{q=-\infty}^{\infty} i^{3q} \left[\prod_{k=1}^{3} J_q(\Lambda_{jk}) \right] \exp(-iq\Delta), \tag{6.4.94}$$

and $\Delta = \Delta_1 + \Delta_2 + \Delta_3$.

SPACE GROUP $P2$

The extension of the Fourier representation of the conditional pdf for the three-phase invariant to this space group was first treated by Shmueli and Weiss (1992). The components of the normalized structure factor for the space group $P2$ are

$$A_m = 2\sum_{j=1}^{N/2} n_j \cos\Theta_{jm} \cos\Psi_{jm} \tag{6.4.95}$$

and

$$B_m = 2\sum_{j=1}^{N/2} n_j \cos\Theta_{jm} \sin\Psi_{jm}, \tag{6.4.96}$$

where $\Theta_{jm} = 2\pi(h_m x_j + l_m z_j)$ and $\Psi_{jm} = 2\pi k_m y_j$ and h, k, l and x, y, x are components of $\mathbf{h}$ and $\mathbf{r}$ respectively. Since we deal with a three-phase invariant, the condition $\mathbf{h}_1 + \mathbf{h}_2 + \mathbf{h}_3 = \mathbf{0}$ must be fulfilled. Therefore the Θ_{jm} and Ψ_{jm} are constrained by

$$\sum_{k=1}^{3} \Theta_{jm} = 0 \text{ and } \sum_{k=1}^{3} \Psi_{jm} = 0.$$

If we substitute eqns (6.4.95) and (6.4.96) into eqn(6.4.89) and perform the usual transformation to polar coordinates, the Fourier coefficient can be rearranged to the form

$$C_{\mathbf{u}} = \prod_{j=1}^{N/2} \left\langle \exp\left[2i \sum_{m=1}^{3} \Lambda_{jm} \cos\Theta_{jm} \cos(\Psi_{jm} - \Delta_m) \right] \right\rangle_{\Theta,\Psi}, \tag{6.4.97}$$

where Λ_{jm} and Δ_m are defined by eqns (6.4.92) and (6.4.84) respectively. The abbreviation

$$\Xi_{jm} \equiv 2\Lambda_{jm} \cos\Theta_{jm}$$

will be made in what follows. We first perform the average over Ψ by making use of the relationship

$$\exp[i\Xi_{jm}\cos(\Psi_{jm} - \Delta_m)] = \sum_{q=-\infty}^{\infty} i^q J_q(\Xi_{jm}) \exp[iq(\Psi_{jm} - \Delta_m)]. \tag{6.4.98}$$

We further substitute the right-hand side of eqn(6.4.98) into eqn(6.4.97), replace Ψ_{j3} by $-\Psi_{j1} - \Psi_{j2}$ and perform the same algebraic steps as in the averaging of the phase factor and calculation of $C_{\mathbf{u}}$ for the space group $P1$. In this way we find the representation

$$C_{\mathbf{u},j} = \sum_{q=-\infty}^{\infty} i^{3q} \exp(-iq\Delta) \left\langle \prod_{m=1}^{3} J_q(\Xi_{jm}) \right\rangle_{\Theta}, \qquad (6.4.99)$$

where $\Delta = \sum_{m=1}^{3} \Delta_m$. We can continue the averaging process by replacing Θ_{j3} in eqn(6.4.99) by $-\Theta_{j1} - \Theta_{j2}$ and making use of the Neumann summation theorem for Bessel functions (Watson 1922; Gradshteyn and Ryzhik 1980). However, a simpler and more symmetric treatment can be based on the fact that the sum of the three Θs for a given atom must be zero. This allows us to introduce a Fourier representation of the periodic Dirac delta function, as already described in subsection 5.3.4 above.

The average on the right-hand side of eqn(6.4.99) then becomes

$$\begin{aligned} \left\langle \prod_{m=1}^{3} J_q(\Xi_{jm}) \right\rangle_{\Theta} &= \frac{1}{4\pi^2} \int_{-\pi}^{\pi} \int_{-\pi}^{\pi} \int_{-\pi}^{\pi} \delta_{2\pi}(\Theta_1 + \Theta_2 + \Theta_3) \\ &\quad \times \prod_{m=1}^{3} J_q(\Xi_{jm}) d\Theta_1 d\Theta_2 d\Theta_3 \\ &= \sum_{s=-\infty}^{\infty} \prod_{m=1}^{3} \left[\frac{1}{\pi} \int_0^{\pi} J_q(\Xi_{jm}) \cos(s\Theta_m) d\Theta_m \right]. \end{aligned} \qquad (6.4.100)$$

The definite integral in the last line of eqn(6.4.100) can be reduced to a product of Bessel functions (Shmueli and Weiss 1992). After doing that, we can summarize the results for the Fourier coefficients for the space groups $P1$ and $P2$ as follows

$$C_{\mathbf{u}}(\Delta) = \sum_{q=-\infty}^{\infty} i^{-q} \exp(-iq\Delta) \mathbf{K}_{\mathbf{u}qj}, \qquad (6.4.101)$$

where

$$\mathbf{K}_{\mathbf{u}qj} = \begin{cases} \prod_{k=1}^{3} J_q(\Lambda_{jk}) & \text{for} P1 \\ \sum_{s=-\infty}^{\infty} \prod_{k=1}^{3} J_{(q+s)/2}(\Lambda_{jk}) J_{(q-s)/2}(\Lambda_{jk}) & \text{for} P2 \end{cases}. \qquad (6.4.102)$$

It has also been shown by Shmueli and Weiss (1992) that the expression for $\mathbf{K}_{\mathbf{u}qj}$ for the space group $P2$ vanishes unless $q + s$ is an even integer.

We are therefore in position to compute the conditional pdf of the three-phase invariant from expressions based on eqns (6.4.87), (6.4.85) and

(6.4.101) with (6.4.102). Such computations have in fact been carried out for the space group $P1$. The results of these calculations will be illustrated in a later section. Before we turn to numerical applications let us examine briefly the reduction of the exact conditional pdfs derived here to the Cochran (1955) result, which was discussed in section 3.5 along with other methods based on the central limit theorem.

6.4.4 *Reduction of the exact result to Cochran's (1955) approximation*

Exact pdfs in structure-factor statistics can usually be reduced to their approximate form based on the central limit theorem, by retaining only the term of lowest order in the expansion of the appropriate characteristic function, and by performing the required Fourier inversion (see, for example, subsection 6.2.2). We follow the same procedure in the case presented in this section. Consider the Fourier integral for the pdf

$$p(\mathbf{E}) = \frac{1}{(2\pi)^6} \int_{-\infty}^{\infty} \cdots \int_{-\infty}^{\infty} C(\boldsymbol{\omega}) \exp(-i\boldsymbol{\omega}^T \mathbf{E}) d^6\boldsymbol{\omega} \tag{6.4.103}$$

and approximate the characteristic function $C(\boldsymbol{\omega})$ by keeping only the lowest-order terms of its real and imaginary parts. For simplicity, we shall do this for the space group $P1$, although it can also be done for $P2$ as well. If only the lowest-order Bessel functions in an atomic contribution are retained, we have from eqn(6.4.94)

$$C_{\mathbf{u},j} \simeq \prod_{k=1}^{3} J_0(n_j\Gamma_k) - 2i \left[\prod_{k=1}^{3} J_1(n_j\Gamma_k)\right] \cos\Delta, \tag{6.4.104}$$

where $\Gamma_k = (\omega_{2k-1}^2 + \omega_{2k}^2)^{1/2}$. When the approximations $J_0 \simeq 1 - x^2/4$ and $J_1 \simeq x/2$ are made, one can find from eqn(6.4.104) the further approximation

$$C_{\mathbf{u},j} \simeq 1 - \frac{n_j^2}{4}(\Gamma_1^2 + \Gamma_2^2 + \Gamma_3^2) - i\frac{n_j^3}{4}\Gamma_1\Gamma_2\Gamma_3 \cos\Delta. \tag{6.4.105}$$

Since

$$\cos\Delta = \cos(\Delta_1 + \Delta_2 + \Delta_3) = \mathbf{V}(\boldsymbol{\omega})/(\Gamma_1\Gamma_2\Gamma_3), \tag{6.4.106}$$

where

$$\mathbf{V}(\boldsymbol{\omega}) = \omega_1\omega_3\omega_5 - \omega_2\omega_4\omega_5 - \omega_2\omega_3\omega_6 - \omega_1\omega_4\omega_6, \tag{6.4.107}$$

the atomic contribution to the characteristic function can be approximated by

$$C_j \simeq 1 - \frac{\boldsymbol{\omega}\cdot\boldsymbol{\omega}}{4} n_j^2 - i\frac{\mathbf{V}(\boldsymbol{\omega})}{4} n_j^3. \tag{6.4.108}$$

If we make use of the approximation $1 - x \simeq \exp(-x)$, we find

$$C(\boldsymbol{\omega}) \simeq \exp\left[-\frac{\boldsymbol{\omega}\cdot\boldsymbol{\omega}}{4} - i\frac{\mathbf{V}(\boldsymbol{\omega})}{4}\sum_{j=1}^{N} n_j^3\right], \tag{6.4.109}$$

which agrees with the expression given by Giacovazzo (1980). One can then follow Giacovazzo's derivation to find that, to the order given, the approximate expression for the conditional pdf of the three-phase invariant is given by

$$p(\Phi|E_1, E_2, E_3) \simeq \frac{1}{2\pi I_0(\kappa)} \exp(\kappa \cos\Phi), \tag{6.4.110}$$

where

$$\begin{aligned} \kappa &= 2\left(\sum_{j=1}^{N} n_j^3\right) E_1E_2E_3 \\ &= 2\sigma_3\sigma_2^{-3/2} E_1E_2E_3 \end{aligned} \tag{6.4.111}$$

with

$$\sigma_n = \sum_{j=1}^{N} f_j^n.$$

Equation (6.4.110) reduces to Cochran's (1955) original result for the equal-atom case, i.e. for $n_1 = \cdots = n_N = N^{-1/2}$, we also have $\sigma_3\sigma_2^{-3/2} = N^{-1/2}$.

6.4.5 *Programming considerations and illustrative examples*

Numerical computation of the conditional pdfs (cpdfs) derived in subsection 6.4.2 presents some difficulties, mainly because of the prohibitive computing times (Shmueli *et al.* 1989a). It is true that six years after the publication of the latter article the attribute 'prohibitive' is less appropriate. Apart from self-evident exploitation of optimization features of the compiler, additional simplifications have been implemented in our program.

The sixfold summation in eqn(6.4.87) is of the form

$$\sum_{u_1}\sum_{v_1}\sum_{u_2}\sum_{v_2}\sum_{u_3}\sum_{v_3} C_{\mathbf{u}}Z_{\mathbf{u}}, \tag{6.4.112}$$

where the indices u_i and v_i range from $-\infty$ to ∞ and each of $C_{\mathbf{u}}$ and $Z_{\mathbf{u}}$ depends on infinite series of Bessel functions. The series converge fairly well but even so, with the upper/lower limit of a single summation being ± 20, and no symmetry being taken advantage of, the summation appears to be a formidable task even for a fast workstation or mainframe (as of 1993). This has been reduced to manageable proportions as follows:

1. The sixfold summation in eqn(6.4.87) is decomposed into partial summations involving five, four, three, two, one and zero summation indices. Each partial summation is computed in a separate subroutine, which increases the program's code but decreases considerably the complexity of branching, and hence results in reduced computing time.
2. The arguments of the Bessel functions appearing in the Fourier coefficient $C_{\mathbf{u}}$ and phase factor $Z_{\mathbf{u}}$ in eqn(6.4.87) depend on the sum of the squares of two summation indices in the sixfold series. The summation can therefore be restricted to positive indices only, as far as the Bessel functions are concerned. However, the phase factors, Δ_k, depend in general on the signs of the Fourier summation indices. The restriction to positive indices (i.e. a decrease of the number of terms by a factor of 64) therefore requires some internal symmetrization of the summation terms. This important time-saving step is explained in some detail in Appendix A to the article by Shmueli *et al.* (1989a).
3. The Fourier summation indices appear in pairs $[(u_1,v_1),(u_2, v_2)$ and $(u_3,v_3)]$ in the arguments of the Bessel functions as well as in the phase factors. If, for example, u_1 and v_1 both run over all the positive integers, we may try to reduce the corresponding pair of summations to the range $u_1 \geq v_1$, which leads to a further reduction of the number of terms by a factor of 8. This is immediate in the case of Bessel functions but calls for some bookkeeping in the case of the phases (see Appendix A in the above-cited paper).
4. Another significant reduction of the computing time results from interchanging equivalent pairs of indices, subject to an internal symmetrization of products of Bessel functions that depend on different magnitudes of normalized structure factors. For example, the Bessel functions entering the Fourier coefficient depend on the pairs (u_1, v_1) and (u_2, v_2) in the same manner, whereas those entering the integrated phase factor $Z_{\mathbf{u}}$ depend on the above two pairs along with the magnitudes E_1 and E_2 respectively. Hence the need for internal symmetrization when interchanging the pairs.
5. It should finally be pointed out, even though this is somewhat trivial, that avoiding repetitive calculations gives rise to major time savings. For example, the required Bessel functions can be computed only once outside the nested summation loop. Similarly, the trigonometric functions depending on integral multiples of the three-phase invariant can also be precomputed once, in the main program. However, table

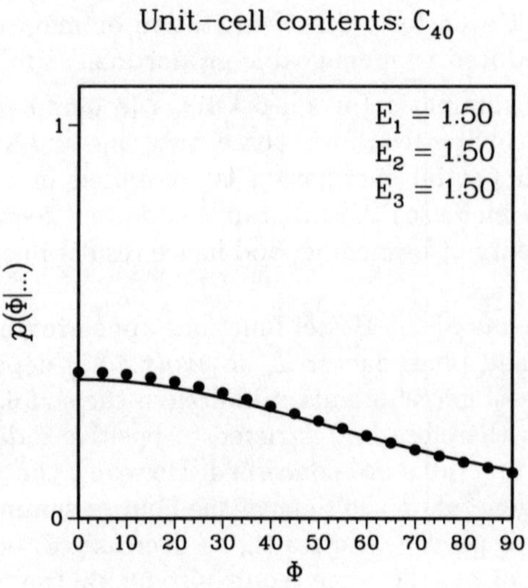

FIG. 6.7. *Cpdfs of Φ for 40 equal atoms and moderately low magnitudes of* $|E|$. The centres of the filled circles are values given by the Fourier cpdf, the solid line corresponds to Cochran's cpdf.

lookup of trigonometric functions was not found to be of great value, presumably because of an improved efficiency of their computation by standard algorithms.

The quantitative aspects of the above simplifications are presented in Appendix A to the paper by Shmueli *et al.* (1989a). It may be interesting to note that after all these simplifications have been correctly implemented, the computing time required for a semi-profile (19 points) of the cpdf of Φ is about 270 seconds on a workstation of HP, series 9000, model 730. This number is quoted as an example of the performance of a fast, modern workstation in 1993 on the problem discussed.

Computations of cpdfs of three-phase invariants have been carried out for a variety of parameters (Shmueli *et al.* 1989b). We illustrate here the results obtained for some cases of interest. Figure 6.7 compares the cpdfs of a three-phase invariant for a $P1$ unit cell containing 40 equal atoms, as obtained from the Fourier representation in eqn(6.4.87) and from the Cochran (1955) expression given in eqn(3.5.110). The invariant is associated with moderately small magnitudes of the normalized structure factors. The two cpdfs shown in Fig. 6.7 are quite close to each other, as expected from the derivation presented in subsection 6.4.3. In fact, in the presence of a sufficient number of equal atoms, and not too strong reflections, the cpdf of the three-phase invariant should tend to agree with Cochran's (1955)

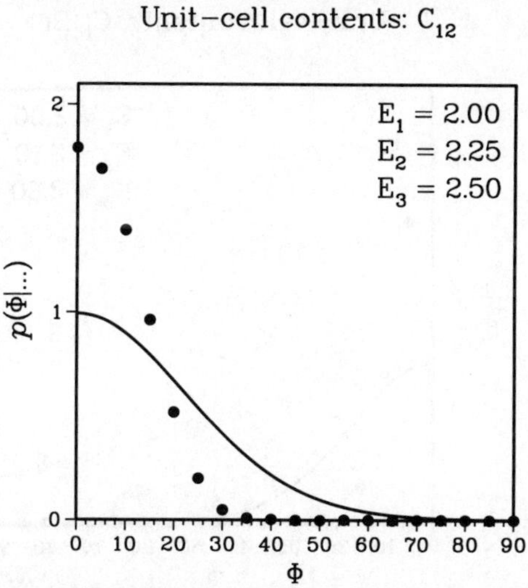

FIG. 6.8. *Cpdfs of* Φ *for a small number of atoms and moderately large magnitudes of* $|E|$. The centres of the filled circles are the values given by the Fourier cpdf, the solid line corresponds to Cochran's cpdf.

equation which is based on the central limit theorem.

As the number of equal atoms in the unit cell decreases, the magnitudes of the Es involved in the structure invariant increase, and the Fourier-based and Cochran cpdfs disagree to a greater extent. An example is shown in Fig. 6.8, for 12 equal atoms in the unit cell and moderately strong $|E|$s. The Fourier-based cpdf is much more sharply peaked than is the Cochran (1955) result. The peak width decreases as the magnitude of the $|E|$s increases. This is consistent with the fact that structures with so poorly populated unit cells can often be solved by deterministic inequality relationships. A qualitatively similar sharpening of the peak corresponding to the Fourier-based cpdf is seen when an outstandingly heavy atom is present, along with not too many light ones. Since the heavy atom dominates the scattering, its presence is analogous to a decrease of the number of atoms. An example of exact and approximate cpdfs based on a moderately heterogeneous asymmetric unit of $P1$ is shown in Fig. 6.9. It is interesting to recall that the routine determination of the phases is generally facilitated by the presence of outstandingly heavy atoms; the significantly decreased variance of the cpdf, in such situations, is concurrent with a decreased phase error and consequently the generation of the starting set becomes more reliable.

It is of interest to calculate the variance of the exact cpdf, and compare it with that of the Cochran (1955) cpdf as given by Karle and Karle (1966)

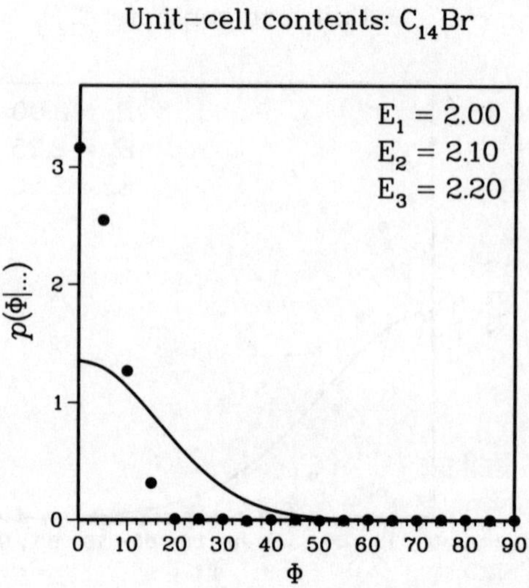

FIG. 6.9. *Cpdfs of* Φ *for a heterogeneous structure and moderately large magnitudes of* $|E|$. The centres of the filled circles are the values given by the Fourier cpdf and the solid line corresponds to Cochran's cpdf.

for the above examples. The conditional variance of Φ, as computed from eqn(6.4.87), is given by

$$\sigma_\Phi^2 = \langle \Phi^2 | E_1, E_2, E_3 \rangle - \langle \Phi | E_1, E_2, E_3 \rangle^2. \tag{6.4.113}$$

It is shown in Appendix A to the paper of Shmueli *et al.* (1989a) that the Φ-dependent part of the summand in eqn(6.4.87) can be written as

$$Z_{\mathbf{u}} = \mathbf{G}_0 + 2\sum_{p=1}^{\infty}(-1)^p \mathbf{G}_{2p}\cos(2p\Delta)\cos(2p\Phi)$$
$$-2i\sum_{p=1}^{\infty}(-1)^p \mathbf{G}_{2p-1}\cos[(2p-1)\Delta]\cos[(2p-1)\Phi], \tag{6.4.114}$$

where

$$\mathbf{G}_q = \prod_{k=1}^{3} J_k(\Omega_k), \tag{6.4.115}$$

$\Delta = \Delta_1 + \Delta_2 + \Delta_3$, Ω_k and Δ_k being given by eqns (6.4.83) and (6.4.84) respectively. It is readily shown that the average $\langle \Phi | E_1, E_2, E_3 \rangle$ vanishes, and a straightforward integration then shows that the conditional variance of Φ is given by

$$\sigma_{\phi}^{2} = \frac{\pi^2}{3} + 8\pi K'\mathcal{R}\sum_{\mathbf{u}} C_{\mathbf{u}} \sum_{p=1}^{\infty}(-1)^p(\mathbf{D}_{2p}\mathbf{G}_{2p} + i\mathbf{D}_{2p-1}\mathbf{G}_{2p-1}), \quad (6.4.116)$$

where $\mathbf{D}_m = \cos(m\Delta)/m^2$. Table 6.1 compares the standard deviations σ_{ex} and σ_{Co}, computed from eqn(6.4.116) and the expression given by Karle and Karle (1966) for the variance of the Cochran cpdf respectively, and expressed in degrees.

Table 6.1

Figure	σ_{ex}	σ_{Co}
6.7	67.5	70.7
6.8	11.8	23.5
6.9	7.8	17.1

The variance again indicates that the exact cpdfs of the three-phase invariant are more sharply peaked than the approximate ones, except when the unit cell of $P1$ contains a moderately large number of equal atoms in which case the cpdfs and their variances are similar.

As can be seen from the mathematical background preceding the derivation of the probability for the Σ_1 relationship, and the reminder on a derivation of n-phase invariants (sections 6.2.1 and 6.4.1), it is formally possible to derive an exact expression for the cpdf of an n-phase invariant, with n greater than 3. However, the computational problems then become rather serious. It would indeed be useful to have exact pdfs for the four-phase invariant (quartet) which are often used in the final stages of the process of phase determination (e.g. Giacovazzo 1993).

6.4.6 *A simple application of complicated formulae*

The derivation and tests of exact cpdfs of three-phase invariants, presented in this section, constitute the most advanced generalization of these important densities published to date. However, the formalism is rather demanding from the computational point of view. Therefore, we do not presently regard these derivations and their results as being directly applicable to the process of phase determination. However, such results could usefully serve as standards with which the various approximate treatments of phase-invariant statistics could be compared. Before carrying out such comparison of theories we decided to first inquire into the feasibility of constructing simple parametric expressions, based on the exact results.

A reasonably large number of computations based on eqn(6.4.87) indicated to us that the exact cpdfs are almost always more sharply peaked than those given by Cochran's (1955) formula, and the latter approximation underestimates the exact cpdf in the interesting region of small values of Φ. This is evident from Figs 6.7, 6.8 and 6.9. Another indication concerned the shape of the cpdfs. Both types of cpdfs, the exact and the approximate,

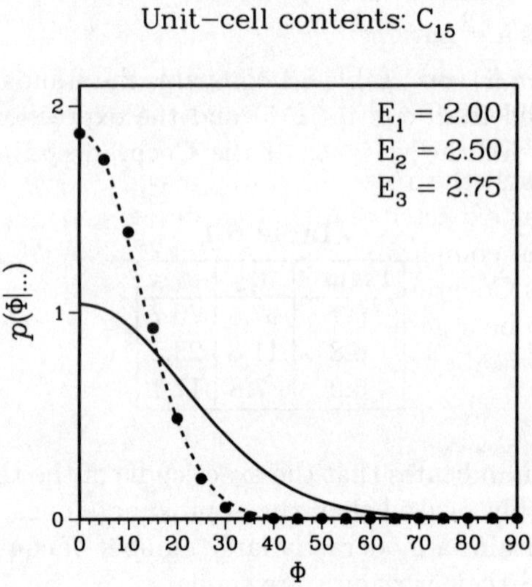

FIG. 6.10. *Cochran (solid), modified kappa (dashed) and exact (•) cpdfs of* Φ. $E_1 = 2.00$, $E_2 = 2.50$, $E_3 = 2.75$, 15 equal atoms in an asymmetric unit.

appeared to have a similar shape and we therefore decided to approximate the exact cpdf by an equation of the same functional form as that given by Cochran (1955). The simplest guess has the form

$$p(\Phi|\kappa') = \frac{1}{2\pi I_0(\kappa')} \exp(\kappa' \cos\Phi), \tag{6.4.117}$$

and rather than fitting κ' to the whole exact profile, we set $p(\Phi|\kappa')$ equal to the exact cpdf for $\Phi = 0$ only. That is, we write

$$p_{exact}(0|E_1, E_2, E_3) = [2\pi I_0(\kappa') \exp(-\kappa')]^{-1} \tag{6.4.118}$$

and solve eqn(6.4.118) for the parameter κ', the numerical value of the left-hand side being computed from the exact formula given in eqn(6.4.87). This simple method, which we have termed 'the modified kappa approximation', produced very encouraging results, as indicated in Fig. 6.10.

Such tests of the modified kappa approximation have been carried out for a variety of choices of parameters (Shmueli and Weiss 1990) and since they were all found to confirm the utility of this simple method we decided to perform a more systematic study of it (Posner, Shmueli and Weiss 1993). We chose the equal-atom case and $|E|$ values ranging independently from 1.25 to 2.75. The values of N were chosen in the range of 15 to 40. The cpdf of Φ given by eqn(6.4.87) was computed at a large number of points of the above range of E_1, E_2, E_3 and N, for values of Φ ranging from 0° to 90°. For each of these cpdfs a κ' parameter was calculated as outlined above and two simple functions of E_1, E_2 and E_3 were fitted to the ratio κ'/κ, where $\kappa = 2E_1E_2E_3/\sqrt{N}$ appears in Cochran's cpdf. The cpdf of Φ in the modified kappa approximation could thus be recalculated and compared with the exact one throughout the above range of E_is and N. The approximate cpdf agrees very well, as expected, with the exact one and the time required for its computation is comparable to that required for obtaining Cochran's (1955) cpdfs. This Cochran-type parametric approximation based on exact results appears to be a good way of utilizing once-computed exact results, which may otherwise be too complex for use in practical applications.

APPENDIX A

We list here for the reader's convenience some known definite integrals which we used in the various derivations presented in the text. The integrals are referred in the text to their source or to this appendix.

$$\int_0^\infty x^{\nu+1}\exp(-\alpha x^2)J_\nu(\beta x)dx = \frac{\beta^\nu}{(2\alpha)^{\nu+1}}\exp\left(-\frac{\beta^2}{4\alpha}\right) \tag{A.0.1}$$

$$\int_0^\infty \exp(-\beta x^2)\cos(bx)dx = \frac{1}{2}\sqrt{\frac{\pi}{\beta}}\exp\left(-\frac{b^2}{4\beta}\right) \tag{A.0.2}$$

$$\int_0^\infty x^{\nu-1}\exp\left(-\mu x^p\right)dx = \frac{1}{p}\mu^{-\frac{\nu}{p}}\Gamma\left(\frac{\nu}{p}\right) \tag{A.0.3}$$

$$\int_0^\infty \frac{\exp(-px)dx}{\sqrt{x(x+a)}} = \exp\left(\frac{ap}{2}\right)K_0\left(\frac{ap}{2}\right) \tag{A.0.4}$$

$$\int_0^\infty x^{\mu-1}\exp(-\alpha x)K_0(\alpha x)dx = \frac{\sqrt{\pi}}{(2\alpha)^\mu}\frac{\Gamma^2(\mu)}{\Gamma(\mu+\frac{1}{2})} \tag{A.0.5}$$

$$\int_{-\infty}^\infty \exp(-x^2)x^m H_n(x)dx = \begin{cases} 0, & n > m \text{ or } n-m = \text{odd} \\ \frac{m!\sqrt{\pi}}{4^p p!}, & m \geq n, p = (m-n)/2 \end{cases} \tag{A.0.6}$$

$$\int_0^\infty \exp(-x)x^m L_n(x)dx = \begin{cases} 0, & n > m \\ \frac{(-1)^n(m!)^2}{n!(m-n)!}, & m \geq n \end{cases} \tag{A.0.7}$$

The integerals (A.0.6) and (A.0.7) can be verified by integration by parts and can be found in Gröbner and Hofreiter (1966). All the other integrals in this appendix can be found in Gradshteyn and Ryzhik (1980).

REFERENCES

1. ABRAMOWITZ, M. AND STEGUN, I.A. (1972). *Handbook of Mathematical Functions.* New York: Dover.
2. BARAKAT, R. (1973). *J.Phys.* A**6**, 796.
3. BARAKAT, R. (1974). *Optica Acta* **21**, 903.
4. BERG, P.W. AND MCGREGOR, J.L. (1966). *Elementary Partial Differential Equations.* San Francisco: Holden-Day.
5. BERTAUT, E.F. (1955). *Acta Cryst.* **8**, 823.
6. BOHL, P. (1909). *Journal für die Reine und Angewandte Mathematik* **135**, 189.
7. BRICOGNE, G. (1984). *Acta Cryst.* A**40**, 410.
8. DE BRUIJN, N.G. (1982). *Asymptotic Methods in Analysis.* New York: Dover.
9. CHARLIER, C.V.L. (1928). *Lunds Univ. Arsskr. N.F. Afd. 2* **24**(8).
10. COCHRAN, W. (1955). *Acta Cryst.* **8**, 473.
11. COCHRAN, W. AND WOOLFSON, M.M. (1955). *Acta Cryst.* **8**, 1.
12. COOLEY, J.W. AND TUKEY, J.W. (1965). *Mathematics of Computation* **19**, 297.
13. COPPENS, P. (1993). In *International Tables for Crystallography, Volume B: Reciprocal Space*, ch. 1.2, ed. U. Shmueli. Dordrecht: Kluwer Academic Publishers.
14. CRAMÉR, H. (1946). *Mathematical Methods of Statistics.* Princeton: Princeton University Press.
15. CRAMÉR, H. (1938). In *Théorie des Probabilités. Actualités Scientifiques et Industrielles*, **736**, 5. Paris: Hermann et Cie.
16. DANIELS, H.E. (1954). *Ann. Math. Stat.* **25**, 631.

17. DANIELS, H.E. (1956). *Biometrika* **43**, 169.

18. DANIELS, H.E. (1980). *Biometrika* **70**, 89.

19. DANIELS, H.E. (1982). *J. Appl. Prob.* **19**, 20.

20. DAVIS, A.W. (1983). In *Encyclopedia of Statistical Sciences*, eds. S. Kotz and N.L. Johnson. New York: Wiley.

21. DAVIS, P.J. AND RABINOVITZ, P. (1967). *Numerical Integration.* London: Blaisdell.

22. FAGGIANI, R., LIPPERT, B. AND LOCK, C.J.L. (1980). *Inorganic Chemistry* **19**, 295.

23. FELLER, W. (1971). *An Introduction to Probability Theory and its Applications, Volume I*, 2nd edn. New York: McGraw-Hill.

24. FOSTER, F. AND HARGREAVES, A. (1963). *Acta Cryst.* **16**, 1124.

25. GIACOVAZZO, C. (1974). *Acta Cryst.* A**30**, 626.

26. GIACOVAZZO, C. (1980). *Direct Methods in Crystallography.* New York: Academic Press.

27. GIACOVAZZO, C. (1993). In *International Tables for Crystallography, Volume B: Reciprocal Space*, ed. U. Shmueli. Dordrecht: Kluwer Academic Publishers.

28. GNEDENKO, B.V. AND KOLMOGOROV, A.N. (1954). *Limit Distributions for Sums of Independent Random Variables.* Cambridge, Mass.: Addison and Wesley.

29. GOOD, I.J. (1961). *Ann. Math. Stat.* 32, 535.

30. GRADSHTEYN, I.S. AND RYZHIK, I.M. (1980). *Tables of Integrals, Series and Products.* New York: Academic Press.

31. GRÖBNER, W. AND HOFREITER, N. (1966). *Integraltafel 2. Bestimmte Integrale.* New York: Springer.

32. HARDY, G.H. AND WRIGHT, E.M. (1979). *Introduction to the Theory of Numbers*, 5th edn. Oxford: University Press.

33. HARKER, D. AND KASPER, J.S. (1948). *Acta Cryst.* **1**, 70.

34. HAUPTMAN, H. (1971). *Zeitschrift für Kristallographie* **134**, 28.

35. HAUPTMAN, H. AND KARLE, J. (1952). *Acta Cryst.* **5**, 48.

36. HAUPTMAN, H. AND KARLE, J. (1953*a*). *Acta Cryst.* **6**, 136.

37. HAUPTMAN, H. AND KARLE, J. (1953*b*). *The Solution of the Phase Problem. I. The Centrosymmetric Crystal.* ACA Monograph No. 3. New York: Polycrystal Book Service.

38. HOWELLS, E.R., PHILLIPS, D.C. AND ROGERS, D. (1950). *Acta Cryst.* **6**, 210.

39. *International Tables for Crystallography* (1983). *Volume A: Space-Group Symmetry*, ed. Th. Hahn. Dordrecht: Reidl.

40. *International Tables for Crystallography* (1992). *Volume C: Mathematical, Physical and Chemical Tables*, ed. A.J.C. Wilson. Dordrecht: Kluwer Academic Publishers.

41. *International Tables for Crystallography* (1993). *Volume B: Reciprocal Space*, ed. U. Shmueli. Dordrecht: Kluwer Academic Publishers.

42. *International Tables for X-ray Crystallography* (1965). *Volume I: Symmetry Groups*, eds. N.F.M. Henry and K. Lonsdale. Birmingham: Kynoch Press.

43. KARLE, J. (1972). *Acta Cryst.* B**28**, 3362.

44. KARLE, J. AND GILARDI, R.D. (1973). *Acta Cryst.* A**29**, 401.

45. KARLE, J. AND HAUPTMAN, H. (1953). *Acta Cryst.* **6**, 131.

46. KARLE, J. AND KARLE, I.L. (1966). *Acta Cryst.* **21**, 849.

47. KHINTCHINE, A. (1949). *Mathematical Foundations of Statistical Mechanics.* New York: Dover.

48. KIEFER, J.E. AND WEISS, G.H. (1984). *AIP Conf. Proc.* **109**, 11.

49. KLUG, A. (1958). *Acta Cryst.* **11**, 515.

50. KLUYVER, J.C. (1906). *Koninklijke Akademie van Wetenschappen te Amsterdam* **8**, 341.

51. Ladd, M.F.C. and Palmer, R.A., eds. (1980). *Theory and Practice of Direct Methods in Crystallography.* New York: Plenum Press.

52. Lighthill, M.J. (1964). *Introduction to Fourier Analysis and Generalised Functions.* Cambridge: University Press.

53. Lipson, H. and Woolfson, M.M. (1952). *Acta Cryst.* **5**, 680.

54. Magnus, W., Oberhettinger, F. and Soni, R.P. (1966). *Formulas and Theorems for the Special Functions of Mathematical Physics.* New York: Springer.

55. Naya, S., Nitta, I. and Oda, T. (1964). *Acta Cryst.* **17**, 421.

56. Naya, S., Nitta, I. and Oda, T. (1965). *Acta Cryst.* **19**, 734.

57. Niederrieter, H. (1992). *Random Number Generation and Quasi-Monte Carlo Methods.* Philadelphia: Society for Industrial and Applied Mathematics.

58. Nigam, G.D. and Wilson, A.J.C. (1980). *Acta Cryst.* A**36**, 832.

59. Pearson, K. (1905) *Nature* **72**, 294.

60. Peschar, R. and Schenk, H. (1986). *Acta Cryst.* A**42**, 309.

61. Peschar, R. and Schenk, H. (1987). *Acta Cryst.* A**43**, 84.

62. Posner, Y., Shmueli, U. and Weiss, G.H. (1993). *Acta Cryst.* A**49**, 260.

63. Press, W.H., Flannery, B.P., Teukolsky, S.A. and Vetterling, W.T. (1990). *Numerical Recipes in Pascal. The Art of Scientific Computing.* Cambridge: University Press.

64. Rabinovich, S., Shmueli, U., Stein, Z., Shashua, R. and Weiss, G.H. (1991*a*). *Acta Cryst.* A**47**, 328.

65. Rabinovich, S., Shmueli, U., Stein, Z., Shashua, R. and Weiss, G.H. (1991*b*). *Acta Cryst.* A**47**, 336.

66. Rogers, D. and Wilson, A.J.C. (1953). *Acta Cryst.* **6**, 439.

67. Sayre, D. (1952). *Acta Cryst.* **5**, 60.

68. Schenk, H., ed. (1990). *Direct Methods of Solving Crystal Structures.* New York: Plenum Press.

69. Shmueli, U. (1982). *Acta Cryst.* A**38**, 362.

70. Shmueli, U. (1993). In *International Tables for Crystallography, Volume B: Reciprocal Space*, ed. U. Shmueli. Dordrecht: Kluwer Academic Publishers.

71. Shmueli, U. and Kaldor, U. (1981). *Acta Cryst.* A**37**, 76.

72. Shmueli, U. and Kaldor, U. (1983). *Acta Cryst.* A**39**, 615.

73. Shmueli, U., Rabinovich, S. and Weiss, G.H. (1989*a*). *Acta Cryst.* A**45**, 361.

74. Shmueli, U., Rabinovich, S. and Weiss, G.H. (1989*b*). *Acta Cryst.* A**45**, 367.

75. Shmueli, U., Rabinovich, S. and Weiss, G.H. (1990). *Acta Cryst.* A**46**, 241.

76. Shmueli, U., Stein, Z. and Weiss, G.H. (1993). *Acta Chimica Hungarica*–Models in Chemistry *130*(2), 261.

77. Shmueli, U. and Weiss, G.H. (1985a). In *Structure and Statistics in Crystallography*, ed. A.J.C. Wilson, pp. 53-66. Guilderland, NY: Adenine Press.

78. Shmueli, U. and Weiss, G.H. (1985b). *Acta Cryst.* A**41**, 401.

79. Shmueli, U. and Weiss, G.H. (1986). *Acta Cryst.* A**42**, 240.

80. Shmueli, U. and Weiss, G.H. (1987). *Acta Cryst.* A**43**, 93.

81. Shmueli, U. and Weiss, G.H. (1988). *Acta Cryst.* A**44**, 413.

82. Shmueli, U. and Weiss, G.H. (1990). In *Direct Methods of Solving Crystal Structures*, ed. H. Schenk. New York: Plenum Press.

83. Shmueli, U. and Weiss, G.H. (1992). *Acta Cryst.* A**48**, 418.

84. Shmueli, U., Weiss, G.H. and Kiefer, J.E. (1985). *Acta Cryst.* A**41**, 55.

85. Shmueli, U., Weiss, G.H., Kiefer, J.E. and Wilson, A.J.C. (1984). *Acta Cryst.* A**40**, 651.

86. Shmueli, U., Weiss, G.H. and Wilson, A.J.C. (1989). *Acta Cryst.* A**45**, 213.

87. Shmueli, U. and Wilson, A.J.C. (1981). *Acta Cryst.* A**37**, 342.

88. Shmueli, U. and Wilson, A.J.C. (1982). In *Crystallographic Statistics: Progress and Problems*, eds. S. Ramaseshan, M.F. Richardson and A.J.C. Wilson. Bangalore: Indian Academy of Sciences.

89. Shmueli, U. and Wilson, A.J.C. (1983). *Acta Cryst.* A**39**, 225.

90. Shohat, J.A. and Tamarkin, J.D. (1943). *The Problem of Moments.* Providence, R.I.: American Mathematical Society.

91. Sierpinski, W. (1910). *Bulletin Internationale de l' Academie des Sciences de Cracovie. Classe des Sciences Mathematiques et Naturelles* Ser A **6**, 9.

92. Sim, G.A. (1958). *Acta Cryst.* **11**, 123.

93. Srinivasan, R. and Parthasarathy, S. (1976). *Some Statistical Applications in X-ray Crystallography.* Oxford: Pergamon Press.

94. Strang, G (1988). *Linear Algebra and its Applications.* 3rd edn. San Diego: Harcourt Brace Jovanovich.

95. Stuart, A. and Ord, J.K. (1987). *Kendall's Advanced Theory of Statistics, Vol. 1.* New York: Oxford University Press.

96. Szegö, G. (1939). *Orthogonal Polynomials*, sections 2.1 – 2.5. New York: American Mathematical Society.

97. Vand, V. and Pepinsky, R. (1953). *The Statistical Approach to X-ray Structure Analysis.* State College: Pennsylvania State Press.

98. Watson, G.N. (1922). *A Treatise on the Theory of Bessel Functions.* Cambridge: University Press.

99. Weiss, G.H. and Kiefer, J.E. (1983). *J. Phys. A* **16**, 489.

100. Weiss, G.H., Shmueli, U., Kiefer, J.E. and Wilson, A.J.C. (1985). In *Structure and Statistics in Crystallography*, ed. A.J.C. Wilson. Guilderland, NY: Adenine Press.

101. Weyl, H. (1916). *Mathematische Annalen* **77**, 313.

102. Wilson, A.J.C. (1942). *Nature (London)* **150**, 151.

103. Wilson, A.J.C. (1949). *Acta Cryst.* **2**, 318.

104. Wilson, A.J.C. (1952). *Nuovo Cimento* **9**, 50.

105. Wilson, A.J.C. (1978). *Acta Cryst.* A**34**, 986.

106. Wilson, A.J.C. (1980). *Acta Cryst.* A**36**, 945.

107. Woolfson, M.M. (1954). *Acta Cryst.* **7**, 61.

108. Woolfson, M.M. (1961). *Direct Methods in Crystallography.* Oxford: University Press.

109. Woolfson, M.M. (1970). *An Introduction to X-ray Crystallography.* Cambridge: University Press.

INDEX